Praise for *Evolutionary Theology*

This excellent and well-written book seeks to provide a coherent understanding of evolutionary theology, what the term "evolutionary theology" means, and how such a theology can refocus questions of original sin, evil, and theodicy. Building on the groundbreaking ideas of Teilhard de Chardin and Alfred North Whitehead, among other contemporary theologians, Michael Abril offers a comprehensive and accessible introduction to the intersection of evolution and theology, laying out various viewpoints and ideas in a simple and concise manner. This book is a welcome addition to the ongoing question of faith and science and will be invaluable for classroom education and adult faith formation.

—Ilia Delio, OSF, Josephine C. Connelly Endowed Chair in Theology, Villanova University

Abril's book offers readable introductions to some well-known Christian theological interpreters of biological evolution. Teachers and students will find the author's succinct summaries of their contributions very thought-provoking.

—John F. Haught, author of *God after Darwin: A Theology of Evolution*

A masterful and panoramic presentation of the philosophical and theological background and outcomes of the idea of evolution.

—Neil Ormerod, honorary professor of theology, Alphacrucis University College, and coauthor of *Creator God, Evolving World*

EVOLUTIONARY THEOLOGY

EVOLUTIONARY THEOLOGY

A
CRITICAL
INTRODUCTION

MICHAEL ANTHONY ABRIL

FORTRESS PRESS
MINNEAPOLIS

EVOLUTIONARY THEOLOGY
A Critical Introduction

Copyright © 2024 Fortress Press, an imprint of 1517 Media. All rights reserved. Except for brief quotations in critical articles or reviews, no part of this book may be reproduced in any manner without prior written permission from the publisher. Email copyright@1517.media or write to Permissions, Fortress Press, PO Box 1209, Minneapolis, MN 55440-1209.

Library of Congress Cataloging-in-Publication Data

Names: Abril, Michael Anthony, author.
Title: Evolutionary theology : a critical introduction / Michael Anthony Abril.
Description: Minneapolis, MN : Fortress Press, [2024] | Includes bibliographical references and index.
Identifiers: LCCN 2024003128 (print) | LCCN 2024003129 (ebook) | ISBN 9781506491639 (print) | ISBN 9781506491646 (ebook)
Subjects: LCSH: Evolution--Religious aspects--Christianity. | Bible and evolution.
Classification: LCC BL263 .A2845 2024 (print) | LCC BL263 (ebook) | DDC 233/.11--dc23/eng/20240326
LC record available at https://lccn.loc.gov/2024003128
LC ebook record available at https://lccn.loc.gov/2024003129

Cover image: Creation of Adam. Hand drawn engraving after Michelangelo, from Marzufello/Shutterstock; engraving of an ape (public domain)
Cover design: Brad Norr

Print ISBN: 978-1-5064-9163-9
eBook ISBN: 978-1-5064-9164-6

To our earliest ancestors
who first experienced the kindling of faith
the fire of divine love
and the embers of unyielding hope

Contents

	Introduction: Theology and Evolution	ix
1.	The Evolution of an Idea: The Symbolic Power of Evolution	1
2.	The Birth of Evolutionary Theology: Pierre Teilhard de Chardin	39
3.	Process Theology and Hegelian Dialectic: Evolution as Co-creation	77
4.	Dialoguing between Doctrine and Science	117
5.	The Evolution of Evil: The Biological Rethinking of Original Sin	147
6.	Future Horizons	181
	Epilogue: An Invitation	207
	Works Cited	217
	Index	237

Introduction

Theology and Evolution

Beauty strikes our hearts even in the most mundane of reflections. Whether we stare out into the abyss of stars or gaze closely at the microcosm of pollinators landing and alighting upon a yellow flower, beauty shines forth in the world around us, a world that Christianity honors by the name of creation. To be creation is to be loved into existence by a Creator.

Yet the way that we gaze at the things of this world does not remain always the same. Today's eyes look through the lens of science. More and more we see the various flowers or the multitude of pollinators as interrelated by evolutionary bonds. Even the stars above are no longer eternal and unchanging but rather the transitory products of eons-long developmental processes.

What does it mean, therefore, to speak of this shifting, changing, ever-developing world by the name of "creation"? Must we abandon such a designation along with outmoded prescientific views, and must we thereby jettison also something of the beauty that holds us captive? Once science has explained away the various processes—atomic, chemical, biological, and physical—that govern the flower and its myriad of buzzing admirers, does creation lose its mystique and cease to call forth wonder?

In truth, for many Christians the world only grows more marvelous as we deepen our understanding of it. Far from destroying the notion of creation, science leads us to a greater appreciation of the world that we believe to be the recipient of God's unfailing love. If evolution is the means by which the Creator establishes this world, then perhaps the science of evolution can tell us something new—something meaningful, beautiful, and profound—about this evolving world. And if this creation bears the marks of its Creator, then perhaps such an exploration can even tell us something about God's very self.

INTRODUCTION

This simple but rich premise forms the basis of *evolutionary theology*, a field of inquiry that seeks to understand the objects of theology (God and all things in relation to God) in light of the insights of evolutionary science.

To some it may seem exotic, superfluous, or even ridiculous to view evolution through a theological lens. What, after all, does God have to do with biology? Did not Charles Darwin develop his theory of natural selection specifically so that we would no longer have to appeal to some divine power in order to explain the natural world? More than that, is there really any longer a place for God within an evolutionary world, or must science lead to a stark *materialism*—the belief that the universe contains no spiritual reality, but only the raw dance of matter and energy, which can be understood through the lens of empirical science alone?

In fact—and not without a dose of irony—whether or not we believe in a deity, God has everything to do with evolution. What many call "evolution" is, after all, more than a mere biological phenomenon. As it carries a weight of meaning that stretches beyond the proper limits of biological science, evolution is in fact a *concept*. From the very beginning, even before Charles Darwin's groundbreaking 1859 publication of *The Origin of Species*, this concept of evolution itself was under development within a cultural and philosophical milieu struggling with the idea of God and the meaning of creation. In various ways, modern thought was becoming infused with a sense of development, which led to the construction of dynamic, progress-based frameworks for understanding the relationship between God and the world.

Darwin, in a sense, wandered into this developing sphere of thought while on his own quest for understanding. Though he was driven first and foremost by his desire to discover, catalog, and understand natural phenomena, these phenomena also became for him a means for approaching larger questions concerning the very nature of life and the possibility of a transcendent God. He observed the preponderance of pain and suffering in the natural world. It seemed altogether unavoidable. He had himself, after all, spent an inordinate amount of his youth shooting birds for pleasure. He came to wonder how the suffering of innocent animals could fit within the traditional view of a just and benevolent God, a God who orders all things toward some higher moral good.

In short, Darwin struggled with the question of *theodicy*: how can evil exist in a universe created by a just God? The goal of moral improvement simply did not satisfy, "for what advantage can there be in the sufferings of millions of the lower animals throughout almost endless time?"[1] Would a loving God create parasites in order to torture their victims?[2] In his quest for theological understanding, Darwin was influenced by John Milton, whose stated intention in writing *Paradise Lost* was to "justifie the wayes of God to men."[3] Darwin sought, in a sense, to hold God accountable for the apparent imperfection of the natural order. Yet in seeking such an account, he did not seek to disown or abandon the existence of God altogether. Rather, he chose to find comfort in an explanation provided by an evolving, developmental view of the world. The natural order came to be as it is not by the ordination of an intelligent divine plan but rather as the slow result of billions of incidental adaptations of creatures to their environments. Suffering, he therefore concluded, stems not from the explicit design of the Creator but rather as the unavoidable baggage of a developing world governed by fixed natural laws.[4]

Darwin's example illustrates how our human urge to understand our ever-changing world has long gone hand in hand with deep questions about God and God's relationship with the world. By invigorating the science of biological evolution, Darwin's discovery had the pronounced side effect of reigniting Western philosophical and theological interest in the interconnection of creatures and their Creator. Though some Christians saw evolution as an attack on traditional values, many others saw it instead as an opportunity for expanding our theological understanding. As early as 1889, the theologian J. R. Illingworth had no qualms about referring to evolution as "an advance in our theological thinking."[5] Thus it was that evolutionary theology, the study of God and the world through an evolutionary lens, was born. This fertile

---

1 Charles Darwin, *The Autobiography of Charles Darwin, 1809–1882*, ed. Nora Barlow (New York: Norton, 1969), 90.

2 Charles Darwin, *The Life and Letters of Charles Darwin*, ed. Francis Darwin, 2nd ed. (London: John Murray, 1887), 2:312; Cornelius G. Hunter, *Darwin's God: Evolution and the Problem of Evil* (Grand Rapids, MI: Brazos, 2001), 12; Ted Peters and Martinez J. Hewlett, *Theological and Scientific Commentary on Darwin's Origin of Species* (Nashville: Abingdon, 2008), 59.

3 John Milton, *Paradise Lost*, bk. 1, line 26; Darwin, *Autobiography*, 85.

4 Charles Darwin, *Darwin and Women: A Selection of Letters*, ed. Samantha Evans (Cambridge: Cambridge University Press, 2017), 176; Cornelius G. Hunter, *Darwin's God*, 11–13.

5 J. R. Illingworth, "The Incarnation in Relation to Development," in *Lux Mundi: A Series of Studies in the Religion of the Incarnation*, ed. Charles Gore, 15th ed. (London: John Murray, 1904), 132.

INTRODUCTION

field of thought has continued to have a profound impact on contemporary Christian viewpoints, both Protestant and Catholic, even into the present day.

Accordingly, in this work we will explore evolutionary theology from its earliest origins into the present day. In particular, we will examine how most contemporary evolutionary theologians have been shaped by one or both of two key influences: Pierre Teilhard de Chardin and Alfred North Whitehead. These influences tend to have a dramatic effect on the scope of evolutionary theology and often shape many of its conclusions. Nevertheless, this field is fertile and varied, such that there is no strict definition or formula. Our goal is not to categorize but to explore. Looking at an array of influential forms of evolutionary theology will provide a glimpse of its possibilities, not only showing how such theology has already been done, but hinting at new ways of integrating theology and evolution in the future.

MOVING BEYOND THE SCIENCE VERSUS RELIGION DEBATE

In order to accomplish our task, we have to move beyond the tired and often simplistic debates between science and religion. Even today, most books on evolution and religion remain stuck within a ceaseless conflict between a narrow, fundamentalist reading of the Bible and an equally narrow materialist view of the world. With all of the futility of a shouting match between Swahili and sign language, both sides of the debate are insulated within completely different worldviews, preventing them from really understanding one another. Worst of all, neither side of this debate can legitimately claim to represent even a plurality of contemporary people, whose openness toward or reservations about the doctrine of evolution often has little to with the dogmatism of either fundamentalism or materialism.

One way or another, whether we accept evolution or challenge it, this has dramatic theological consequences that must be taken into consideration. Evolution has met with religious resistance not simply because it puts into question a literal reading of the Bible, but even more because it forces us to rethink key theological assumptions about the meaning of the world, the dignity of humanity, the origins of good and evil, and the significance of history, technology, and the future.

Some Christians have rejected evolution on these grounds, but many others have sought to forge fruitful relationships between science and theology.

Thus, evolutionary theology arose as the hope that the idea of evolution might provide insight into a deeper theological truth, forming a cohesive and responsive worldview that could incorporate God and the world into one harmonious perspective. This is the ongoing story of evolutionary theology. While often controversial in its own right, particular forms of evolutionary theology have also had a dramatic effect on the way many Christians approach issues of science, faith, and the future.

INVESTIGATING A CONCEPTUAL HERITAGE

While many readers may find the notion of evolutionary theology somewhat novel, in fact what we are exploring here is in many ways integral to contemporary thinking. I am not in fact introducing a complete stranger. Rather, this is an encounter with someone whom you have already met. Learning to recognize the conceptual role of evolution today opens our eyes to see how this concept is already deeply enmeshed within our everyday experiences. It is that friend who always seems to stand at the periphery of the camera, always a part of our vision even if seldom in focus.

Because of this, though we will often discuss matters of history, we will find that this history is more properly understood as *heritage*. This is not a book about what some dead men happened to believe over the past hundred and fifty years. Rather, it is a book about how the things we casually believe, the way we view our world, and many of the actions that stem from this worldview are intimately shaped by a complex intellectual, cultural, and even theological heritage formed in conversation with the scientific doctrine of evolution.

We are shaped by conceptual factors even before we come to recognize and understand them. For example, our thoughts, language, and actions are often haunted by categories borrowed from Freudian psychoanalysis. Modern people think in terms of repression, of projection, and of subconscious motives, even without having read a page of Sigmund Freud's works.

In much the same way, ideas of evolutionary development frequently shape how contemporary people think, believe, and act. As Mary Midgley points out, evolution "influences not just our thought, but our feelings and actions too, in a way which goes far beyond its official function as a biological theory."[6]

---

6 Midgley, *Evolution as a Religion: Strange Hopes and Stranger Fears*, rev. ed. (London: Routledge, 1985), 33.

INTRODUCTION

At home with viewing the world as enmeshed in constant flux, we tend to see our own human story—both as individuals and as a species—as a tale of gradual adaptation and ongoing progress.

Just as it is easy for advertisers to tap into men's innate desire for women in order to sell a diamond, a fragrance, or an outfit, so also it has become easy to use our contemporary assumptions about nature, evolution, and meaning to promote a wide variety of products, ideas, and even lifestyles. Take for example today's most popular dieting techniques, which are often marketed as being calculated to suit evolution's design for human nature. "We are not living as we were built to live," argues one Paleo diet evangelist.[7] Humans first evolved in an environment marked by frequent periods of starvation, sporadic episodes of fight or flight, and an eclectic menu consisting of whatever "you (or someone else) can either pick or catch and kill."[8] Thus, he argues, the path to ideal health demands periodic fasting, random bursts of energetic self-exertion (not regimented exercise), and a general aversion to any food that was not widely available to our ancestors roughly forty thousand years ago. Nature, like a mother, knows best. For such a view, evolutionary science uncovers the truth about who the human animal is and how one should live in relation to this world.

If anything, such popular evolutionary rhetoric shows that in practice it is far from easy to separate out a purely scientific perspective from our underlying ideological or even theological commitments. As John Haught points out, even dedicated atheists and materialists such as Richard Dawkins and Daniel Dennett have tended time and again to slip back into the realm of theology.[9] One can say scientifically that evolution represents such a complete theory that it provides no space for any special divine intervention within its own inner workings. However, as soon as one says that evolution actually rules out or disproves God or otherwise reshapes our understanding of God, we have crossed back over into the realm of theology and have begun to make claims that transcend the scope of empirical science.

If we maintain that science and faith are two entirely autonomous and mutually exclusive spheres, such that one has nothing to say to the other, then

---

7 Arthur De Vany, *The New Evolution Diet: What Our Paleolithic Ancestors Can Teach Us about Weight Loss, Fitness, and Aging* (Emmaus, PA: Rodale, 2011), 6–7.

8 De Vany, 40.

9 John F. Haught, *Making Sense of Evolution: Darwin, God, and the Drama of Life* (Louisville, KY: Westminster John Knox, 2010), 18.

INTRODUCTION

our own failures to maintain a strict separation of disciplines must repeatedly place these two in conflict to the detriment of both realms.

In contrast to this, a long tradition of theologians has taken an entirely different approach. Rather than beginning with the assumption that faith and reason must be incompatible, they have seen the ongoing dialogue between evolution and theology as a mutually constructive relationship.

Some thinkers, like Pierre Teilhard de Chardin, have seen evolution as a beautiful matrix of divine meaning, leading to the incarnation of Christ and ultimately pointing onward toward the culmination of God's perfect plan throughout the world. God illuminates the meaning of evolution. Others, going further, see evolution not merely as a story of the world's development but as a drama that involves God's very own self-development in and through the world. Evolution illuminates the meaning of God. Recent years have only intensified efforts in evolutionary theology, especially as new insights in biology, ecology, and psychology have reframed many of the important questions that shape this kind of theology as a discipline.

My purpose in this book is to explore the origins and shape of evolutionary theology in order to provide critical insight into its significance today. For this reason, we must delve into a theological conversation at the very origin of the theory of evolution, an ongoing conversation that has continued to evolve and to adapt to new questions, challenges, evidence, and ideas until the present day. This book thus serves an important function not only with regard to this limited field but to the broader realm of contemporary theology. Like the theory of evolution itself, evolutionary theology is an integral and inescapable part of our contemporary world. To some extent, speaking the Gospel in our present-day context requires familiarity with the various forms of evolutionary theology and the key strategies that they utilize to respond to important questions arising out of this context.

THE IMPORTANCE OF A CRITICAL INTRODUCTION

In view of this significance, formulating a critical introduction to evolutionary theology represents a bold and daunting task. A simplistic, one-sided presentation of evolution's theological relevance cannot produce the kind of critical comprehension required by genuine forward progress. Without a sense of evolutionary theology's historical development and a firm understanding of its scientific, philosophical, and theological bases, scholars risk accomplishing little more than walking in circles. Is it any surprise that certain twenty-first century theologians propose as groundbreaking and new some of the same basic doctrines that James Orr critiqued back in 1903?[10] A mere theological absorption of cutting-edge scientific theories or discoveries will have little lasting impact on its own. It must be contextualized within the fuller conversation between theological tradition and modern evolutionary science.

Despite this need, works of evolutionary theology seldom provide a productive understanding of this vital context. The richness and diversity of the field of evolutionary theology has produced and continues to produce a broad array of lectures, essays, and books. One can easily find a wealth of works that propose constructive arguments, formulate conceptual frameworks, offer interpretations of the Bible, investigate the theological significance of specific scientific ideas, or make core concepts accessible on the popular level. Nevertheless, these contributions are seldom rooted in anything more than a cursory, one-dimensional exploration of the conceptual bases, history, and development of evolutionary theology as a field. Most often, even longer works provide only enough background material to introduce the constructive arguments of the text.

The present work thus seeks to provide the necessary context for a fuller and more fruitful understanding of contemporary evolutionary theology. As an introduction, its primary concern is to lay out various viewpoints and ideas as plainly and accessibly as possible without oversimplification. This means that critique takes a back seat to explanation. While I do offer some critiques, especially in the concluding chapter, I try for the most part to allow the reader to formulate personal conclusions from the information provided. As Leo Foley points out, "Criticism is not only negative."[11] The critical aspect of this book lies in its incisive investigation into the subject matter.

Despite important commonalities among evolutionary theologians, the remarkable diversity of the field undermines any attempt to oversimplify it and its relationship to Christian tradition. In essence, all Christian theology is a reflection upon tradition. Even a militantly post-Christian feminist theology,

---

10 James Orr, *God's Image in Man and Its Defacement in the Light of Modern Denials* (London: Hodder and Stoughton, 1905), 206–207.

11 Leo Albert Foley, *A Critique of the Philosophy of Being of Alfred North Whitehead in the Light of Thomistic Philosophy* (Washington, D.C.: Catholic University of America Press, 1946), x.

by diagnosing Christian tradition as irredeemably contaminated by patriarchy, takes thereby a stand relative to this tradition. Evolutionary theology is no different. Some approaches, like that of Pierre Teilhard de Chardin, begin out of a professed concern for Christian tradition and seek to deepen the conversation between that tradition and contemporary science. In contrast, process-based approaches rooted in the metaphysics of Alfred North Whitehead have fostered a dramatically different approach. Whitehead's approach is favored among many who seek to criticize and overturn traditional views on God, God's relationship to the world, and the function of the church in society. Adherents see a process-based approach as providing better access to Christianity's core faith by stripping away problematic metaphysical concerns rooted in Greek philosophy.

In effect, evolutionary theology concerns not merely the biological evolution of the human race but also the evolution of *Christianity* itself. If Christianity has developed in its beliefs, concepts, language, and practices over the past two thousand years—and indeed, it has—then to what extent might this ongoing history of development align with the general evolutionary viewpoint rooted in modern science? Evolutionary theology thus brings to the fore vital, fundamental questions about Christian tradition, the same questions to which Christians have sought deeper answers since the dawn of the modern era: To what extent must Christian tradition be held accountable to contemporary attitudes, morals, and ideas? What ultimately is perennial—what *is* the Gospel at its finest—and what represents a mere transitory development, a relic of a particular time and place, which might ultimately be stripped away, having lost its relevance at the passing of the original circumstances that fostered its development?

This is not to say that evolutionary theology's primary function is to justify a more progressive, modern, or reductionistic stance on tradition. Quite the contrary: a majority of Christian theologians believe that Christianity has, in some sense, developed. The principal debate within Catholicism over the significance of the Second Vatican Council, for example, is not between those who accept development and those who reject it, but rather between those who see its development as an organic and providential deepening of what came before and those who see the council as a dramatic reversal or nullification of outmoded views and attitudes.

Evolutionary theology, likewise, incorporates diverse and even diametrically opposed views on the history of the development of Christian doctrine. Yet it provides also an important angle for approaching this issue precisely because it concerns itself with conceptualizing that development. Is Teilhard correct to see Christianity's progress as fostered by the work of the scientist, whose occupation takes on a certain religious (and even priestly) significance? Or do we adopt Whitehead's essentially Hegelian, dialectical view of history, wherein a negative, rebellious counterthrust is necessary to overturn the stagnant dogmatism of the status quo? Is the development of tradition a peaceful and cooperative process or else the catastrophic but necessary destruction of our conceptual idols?

OUTLINE OF THE WORK

Our investigation into evolutionary theology will cover an array of figures, books, and ideas, but of course we cannot cover everyone and everything of significance without sacrificing readability. While there are many theologians who emphasize some sort of developmental perspective, for our purposes we will focus on theologians who explicitly see the biological science of evolution as informative for our theological perspective. Even such a narrow focus, however, will require us to branch out in various directions. Whitehead, for example, does not focus heavily on biological evolution, and yet he is a vital influence behind many others who do.

Our first destination on this journey of discovery will take us back to the formulation of the scientific doctrine of evolution. We will explore how philosophical and theological ideas of development both added to and subverted scientific discovery so that what emerged was not merely a scientific notion of evolution but also various ideological forms of evolution*ism*. As evolution comes to challenge humanity's privileged place in the cosmos, various thinkers will turn to evolutionism for new ways of conceptualizing humanity's significance.

From there, we will examine the most significant influences behind contemporary evolutionary theology. In chapter 2, we will see how key thinkers responded to the challenges of evolution by revamping natural theology in light of philosophical ideas of development. They develop key insights that coalesce within the work of Pierre Teilhard de Chardin, whose broad influence makes him effectively the father of evolutionary theology. In chapter 3, we examine the ideas of another central figure, Alfred North Whitehead. We explore Whitehead's relation to Georg Wilhelm Friedrich Hegel in order to illuminate influential aspects of his thought.

This foundational approach establishes a basis upon which we can explore the impact of evolutionary theology on key dogmatic questions. In chapter 4, we look at the vital questions posed by the Catholic encyclical *Humani generis* and the formative responses of Karl Rahner. Chapter 5 then focuses more specifically on the doctrine of original sin, examining several ways in which theologians have sought to reconceptualize the doctrine in light of evolutionary science.

As Jerry Korsmeyer points out, evolutionary theology seeks not only to aid our understanding of the past but also to "assist humanity in dealing with its future."[12] Accordingly, chapter 6 explores many of the ways in which evolutionary theologians conceptualize the coming future. It explores avenues such as moral and cultural development, technological exploration, and ecological action.

It is my genuine hope that this work will not merely catalog existing forays in evolutionary theology but will sustain new and improved entries into the field. If nothing else, I hope the reader will discover herein a sense of the wonder, the mystery, and the excitement of peering into the mystery of God and creation through the lens of an evolving universe.

ACKNOWLEDGMENTS

Before we proceed, I would like to thank those whose assistance has been vital to this undertaking.

First and foremost, thank you to my wife Danielle, whose example of Christian charity both inspires my work and allows me the space to complete it. Thank you to my son and daughters for bearing with me and always greeting me with love and smiles. Thank you to all of those pioneering scholars in evolutionary theology for inspiring this study.

I am also indebted to the many persons who have read and reviewed drafts of this work, especially among my students. Thank you in particular to my graduate assistants J. Peter Swindeman, Rafael Piña Valdez, and Madison Williams for their care in research, their companionship in discussion, and their diligence in reading drafts. Thank you to Fortress Press for bringing this book to the public and especially to my editor Ryan Hemmer.

Last but never least, if this study says anything true or beautiful, then I must credit the most gratuitous and awe-inspiring grace of the Holy Spirit.

---

12 Jerry D. Korsmeyer, *Evolution and Eden: Balancing Original Sin and Contemporary Science* (New York: Paulist, 1998), 126.

CHAPTER ONE

The Evolution of an Idea
The Symbolic Power of Evolution

Few ideas have had such a monumental impact on the history of human thought as the theory of evolution. Whether we choose to accept evolution or debate it, on some level we are all still faced with the necessity of coming to terms with it. J. R. Illingworth's insightful 1889 remarks are just as relevant today: "Evolution is in the air. It is the category of the age. . . . We cannot place ourselves outside it, or limit the scope of its operation."[1] Our thoughts are deeply shaped by the evolutionary paradigm. "Organisms, nations, languages, institutions, customs, creeds, have all come to be regarded in the light of their development, and we feel that to understand what a thing really is, we must examine how it came to be." We are predisposed toward seeing the world and its parts as moving in stages from simple to more complex, from lesser to greater, from one state of affairs to another.

No doubt the rapid advancement of technology in the past few decades has only served to solidify this perspective. Cartoons have evolved from pictures hand-drawn and painted to complex 3D computer-generated models shaded with realistic textures and lighting effects. Telephones have evolved from clunky wall-mounted rotary-dialed fixtures to tiny, portable all-in-one entertainment systems. Our very sense of time is shaped by such developments; one can have a reliable sense of "back before HDTV," without being able to name the year one first bought a high-definition set. "The incredible technological changes over the last century separate us from our past more

---

1 Illingworth, "Incarnation," 132.

effectively than the years," writes a pair of anthropologists.[2] In short, we experience the world as evolving on many levels, and this experience is so tangible and vivid that it easily becomes a major basis by which we interpret our world. Even without us realizing it, we often rely implicitly upon one or another form of the theory of evolution in order to understand our very experiences of time, life, and meaning.

Yet ironically, evolution obtains a significant part of this interpretive influence from the fact that our conceptions of biological or cosmological development often have less to do with biology or physics than with philosophy and theology. Even within the writings of some prominent scientists one sometimes finds at work not merely the scientific theory of evolution but also a philosophical, semiological, or even theological *concept* of evolut*ism*. In short, evolution is so inspiring an idea precisely because in the first place it has always been received culturally, philosophically, and religiously as *more* than a mere statement of objective fact. This evolutionism is a grand, all-encompassing idea—a *metanarrative*.[3] It is a story that purports to explain all others, giving order and meaning to all reality by its claim to see the forest beyond the trees. "History is the story of the world's progress from lesser to more, from simple to complex, from chaos to harmony"—so says one version of the evolutionism metanarrative.

To understand how evolutionism has taken hold and how this has fed into the rise of evolutionary theology, we must look back *behind* the origins of evolution as a scientific theory. As we shall see, nineteenth-century scientists put forth and planted their pivotal contributions within an intellectual field already made fertile by a wealth of philosophical ideas. The theory of evolution had a monumental impact precisely because it touched upon deep-seated modern inclinations toward progress and development. Yet as a result, many thinkers took up the science of evolution as confirmation of their own philosophical, political, or theological views. They intermingled the authority of empirical science with the imaginative beauty of a grand idea. By thus attaching external, conceptual baggage to the theory of evolution, even many early

---

2 Milford H. Wolpoff and Rachel Caspari, *Race and Human Evolution* (New York: Simon and Schuster, 1997), 50.

3 For a detailed examination of the difference between scientific evolution and ideological evolutionism, see Raymond J. Nogar, *The Wisdom of Evolution* (Garden City, NJ: Doubleday, 1963), 242–51. Nogar defines evolutionism more narrowly, however, as an ideology of evolution that has achieved exclusive dominance over life and values.

evolutionists crossed the critical line between evolution and evolutionism. This conceptual transgression of sorts will spark the imagination of many religious thinkers, who come to think of the world in terms of evolutionary development. At the same time, challenging traditional assumptions about the meaning and centrality of the human person, the advent of evolutionary thinking will force many such theologians to re-root these theological ideas within the framework of evolutionism. Hence evolutionary theology will be born out of and as a critique of the evolutionary metanarratives of the late nineteenth century.

THE EVOLUTION OF DARWIN'S CONCEPTUAL CONTEXT

Popular history focuses on Charles Darwin (1809–1882) as the father of the theory of evolution. Scientists know well, however, that Darwin did not happen upon this theory on his own. In the late eighteenth and early nineteenth centuries, scientific ideas about biological change were already taking shape among scholarly minds. A long history of scientific successes and failures met with an ever-growing catalog of empirical observations to propel scientists toward considering how the biological world as we know it might be shaped by factors of time, environment, and heredity. The fertility of this nineteenth-century intellectual context actually led *two* scientists to formulate the basis of the modern scientific theory of evolution at the same time: Charles Darwin and his contemporary Alfred Russel Wallace (1823–1913). Yet it was not evolution *as such* that Darwin and Wallace discovered, as though no one else had considered that organisms might change over time. Rather, they performed the critical service of uncovering and unpacking the key environmental and systematic mechanism by which such change occurs: *natural selection*.

The scientific theory of evolution is a distinctively modern contribution, and as such it could not have come about without a long list of scientific and philosophical developments that led Darwin and Wallace to view the natural world in a new way. These developments, moreover, were born out of the complex intellectual movements that led forth out of the Middle Ages, through the Reformation, and into Enlightenment and modernity as such. A fuller picture of the origins of the theory of evolution thus requires a glance at some of the earlier philosophical, theological, and cultural influences that continued to shape much of the mindset of the nineteenth century. The theory of natural selection shed light on the biological process of *transmutation*—the production of one species from another—but broader ideas of development preceded Darwin and Wallace and informed the way in which many people read and understood their works. These ideas of development were forms of *evolutionism*—evolution not as a scientific theory but rather as a philosophical metanarrative. To some extent, the popularity of evolutionism aided the reception of Darwin's theory. However, it also laid a trap. This complex conceptual background ensured that Darwin's idea would also be co-opted and misunderstood, manipulated into an empirical justification for prior philosophical metanarratives that sought to explain far more than mere biology.

We cannot hope to explore a full history of the development of the theory of evolution in this limited space. In particular, we cannot hope to do justice to the scientific discoveries of the peers and immediate predecessors of Darwin and Wallace. Rather, in order to provide a more conceptual and thematic sketch of their context, it will help to point to a few significant moments that illuminate the origins of evolutionism as an idea and to see how its development both differed from and influenced the discovery of evolution as a scientific theory.

The Possibility of a Totalizing Worldview: Plato and Aristotle

The true story of evolutionism—of the philosophical perspective that sees the world as continuously developing—begins with Plato (ca. 428–347 BCE).[4]

---

4 Some mention should be made of the fascinating and fragmentary ideas of pre-Socratic philosophers, whose comments can seem, at first glance, to foreshadow the science of evolution. Anaximander (sixth century BCE) not only correctly theorized the marine origin of land animals but also posited that humans must have been born from another animal. A century later, Empedocles sang a story about the origin of species that echoes the idea of natural selection. In his song, animal life began with the chaotic, random combinations of organs (e.g., an ox head with a human torso). By chance, among such strange monstrosities there occurred also appropriate combinations (e.g., a human head with a human torso). The chaotic miscombinations failed because of their inadequacy and thus perished. As a result, the ordered, well-adapted combinations came to dominate the world as we know it. Empedocles, *The Poem of Empedocles: A Text and Translation with an Introduction*, trans. Brad Inwood, rev. ed. (Toronto: University of Toronto Press, 2001), 124. Despite appearances, however, such pre-Socratic hints are really only amusing coincidences. Counted among the many other factually false suggestions that early

Both in themselves and through Plato's student Aristotle (384–322 BCE), Plato's views form the basis of a broad range of Western thought, Christian and pagan. This deeply Platonized Christian philosophical milieu, spurred by new ideas and important cultural and historical shifts, gave birth to the Enlightenment and modern thought. Thus even as modern thinkers came to challenge traditional Christian thinking, they often did so on the basis of Platonic, Aristotelian, and Neoplatonic forms of thought (ca. 244–529 CE). Such thinking thus served also as the inescapable background against which nineteenth-century thinkers questioned the transformation of biological species.

Historical studies of the emergence of Darwinism often exclusively treat our Platonic and Aristotelian heritage as a problem or obstacle, which sciences had to overcome in order for the concept of natural selection to flourish. This is at least half true. To some extent, both figures contributed to what is termed "typological thinking," the tendency to consider particular species as representing static, unchangeable, and well-defined types. For Plato, who saw the physical world as nothing more than a shadow of a more real and fundamental world of unchanging Ideas, the true definition of any particular species is essentially written always and forever within the very fabric of reality. If a particular cat is a mere manifestation of the eternal Idea of "Cat," then apparent variations among cat populations are mere unimportant accidents. The more perfect an individual cat, the better it approximates the original Idea of "Cat" after which it is patterned. Individual cats will come and go, but the type or category of animal called "Cat" will always exist, since the beauty of this world consists in its reflecting (however imperfectly) the multiplicity of the world Ideas.[5] Within such a framework, the arrival of an altogether new species is as difficult to imagine as the possibility of extinction.

Aristotelian philosophy, while eschewing Plato's reliance upon an immaterial sphere of Ideas, did not altogether nullify this typological thinking. His contribution is mixed. On the one hand, Aristotle actually provided a basis for questioning the typological framework of Plato. By analyzing living creatures

---

philosophers made, they have little real significance. See Gould, *Ever Since Darwin*, 202. Read in context, Empedocles's ideas about the development of the world through the swirling forces of love and strife have more in common with the philosophical views of Denis Diderot (1713–84) and F. W. J. Schelling than with the scientific theory of Charles Darwin. See Peter J. Bowler, *Evolution: The History of an Idea*, 25th anniv. ed. (Berkeley: University of California Press, 2009), 19–20, 82.

5   See Plotinus, *Enneads* II, 1.

according to their observable characteristics, Aristotle recognized that while different species share many characteristics, there is no way to classify them on a simple line from least to greatest. The natural world is a complex web of similarities and differences. Any single creature tends to evade clear-cut definition. He even recognized that traits can vary among individuals of a species and that such variations can sometimes be inherited by offspring.[6]

On the other hand, the subtlety of Aristotle's views could not of itself overturn the trend of typological thinking. In fact, this tendency became exaggerated via the rise of scientific systems of classification near the end of the seventeenth century, which relied on the assumption that nature is inherently stable and unchanging.[7] In order to group creatures according to well-defined species, early taxonomists downplayed differences among individuals or populations within a species. Variations among particular cats were insignificant, for example, since they could not be used to distinguish cats from dogs.

Nevertheless, Plato and Aristotle also shaped Western thought in broad ways that would ultimately lead nineteenth- and twentieth-century thinkers to embrace the idea of biological evolution. In fact, notwithstanding their typological basis, the formation of taxonomical systems in the late seventeenth century proved an important step toward the idea of biological evolution. Plato and Aristotle provided a key assumption that continued to shape Western ideas about biology even after typological thinking faded into the growing suggestion that species might, in fact, be capable of change. Specifically, they provided a sense of the cohesive unity of the world, an interconnectedness that allowed later thinkers to conceive of species as slight modifications of one another. This Greek philosophical standpoint of unity played a key role in the Christian view of the world as purposeful, and yet it became also, ironically, the basis of early mechanistic views that would contribute to the rise of evolutionary science.

For Plato, the world exhibits the beauty and cohesion of the realm of Ideas by containing within itself each and every creature in its individual uniqueness. The world is itself a perfect organism, a "god," precisely because the diversity of worldly creatures mirrors the plenitude of the immaterial

---

6 For example, Aristotle, *History of Animals*, trans. D.M. Balme, vol. 3 (Cambridge, MA: Harvard University Press, 1991), 455. For a detailed analysis, see Fran O'Rourke, "Aristotle and the Metaphysics of Evolution," *Review of Metaphysics* 58, no. 1 (2004): 3–59.

7 Bowler, *Evolution*, 45.

realm of Ideas.[8] The differences among creatures—between birds and beasts, humans and apes, and so forth—thus confirm the cohesive unity of the world. It could be no other way. Whatever god fashioned the world was forced to do so by an inner necessity—by his own essential relation to the Ideas. A good god must embody or express this goodness through the creation of a unified and cohesive world, which alone could serve as a suitable (though constitutively imperfect) reflection of the perfect divine Ideas. The assortment of distinct species, for example, makes the world like a beautiful painting, which is beautiful not because of the individual dabs of paint but because of the arrangement of the colors as a whole.

This affirmative take on the differences among species led to a key development in Aristotle, who saw such differences as evincing the fundamental complexity of the natural world. Any attempt to classify creatures based on their properties ultimately reveals gradual scales of difference rather than strict lines of demarcation. He remarked, "Nature proceeds from the inanimate to the animals by such small steps that, because of the continuity, we fail to see to which side the boundary and the middle between them belongs."[9] Even the seemingly binary division between plants and animals is blurred by the existence of immobile, plantlike animals such as the sponge. All creatures are interrelated and share similarities as well as differences. Aristotle did not go so far as to claim that any species developed out of another. Nevertheless, by highlighting the interrelations of creatures by their measurable characteristics, he provided a means of classifying them, at least logically, according to their similarities and differences.

Such classification by characteristics need not be hierarchical; it highlights the connections among creatures without necessarily seeing each one as better or worse, higher or lower, than others. Aristotle made no attempt to rank specific characteristics such as horns and scales.

Still, hierarchy was central to his master Plato, and Aristotle did not strictly renounce it. To some extent, he did illustrate the relative superiority of humans over other animals and animals over plants according to how intensely each kind of creature exhibits qualities such as sensitivity, freedom of mobility, and intelligence. Such hierarchical thinking—already implicit within

---

8 Plato, *Timaeus*, 30–31 (Stephanus pagination). On the relation between Plato's gods and the Ideas, see Étienne Gilson, *God and Philosophy* (New Haven, CT: Yale University Press, 1941), 26–29.

9 Aristotle, *History of Animals*, 3:60–63.

Plato—would shape the medieval mindset and eventually become known by the Latin phrase *scala naturae*, "the ladder of nature." To be fair, Aristotle's version of the ladder was more implicit than explicit and more bumpy than straight. Humanity's relative superiority did not make all animal characteristics categorically inferior. He recognized, for example, that many animals surpass humans in their sense of smell. Moreover, what counted as a perfection for a human could be a defect for another creature; a bird would not benefit from humanity's forward-facing knees.

Nevertheless, the hierarchical tendency of the *scala naturae* came to surpass Aristotle's own intentions. Building upon a mix of the work of Aristotle and Plato, Stoicism, and Neoplatonism, Christians such as Nemesius of Emesa (late fourth century CE) viewed all creatures as interrelated by an overarching ontological staircase, with humans standing closer to the top than most others. We are the most noble among embodied creatures because we are also spiritual beings and thus capable of knowing and communing with transcendent, spiritual realities. This makes us also an important link in this "Great Chain of Being"—as Alfred O. Lovejoy famously termed it[10]—for we implicitly unite embodied reality (down to the merest inanimate mineral) with spiritual reality (up to the very perfection of being: God).[11] Although in Platonic fashion each creature could be seen as having its own purpose and validity, the *scala* also provided a sense in which so-called lesser organisms could be seen as directed toward higher beings and humanity in particular. By reaching into the spiritual realm, humanity attains to a cosmic fulfillment that reflects back upon the rest of the bodily, material world.

The Christian Historical Emphasis: Augustine

In this way, the contributions of Plato and Aristotle set the tone in which Western philosophers and Christian theologians would discuss the natural world for roughly the next two thousand years. By understanding the world as a cohesive unity and ordering this unity hierarchically in terms of the *scala naturae*, Christian tradition viewed the natural world primarily in terms of its

---

10 Arthur O. Lovejoy, *The Great Chain of Being: A Study of the History of an Idea* (Cambridge, MA: Harvard University Press, 1936), 59.

11 Nemesius, "A Treatise on the Nature of Man," in *Cyril of Jerusalem and Nemesius of Emesa*, ed. William Telfer (Philadelphia: Westminster, 1955), 228–35.

cohesive *meaning*. The created world fits into the divine plan that culminated with the incarnation. Through its interrelation with Christian doctrine as a whole, this view thus served as a *metanarrative*: a grand, cosmic vision of the universe that explains its origin, meaning, and purpose.

This fit well within the medieval Christian milieu, which inherently saw the world as a manifestation of divine beneficence, craftsmanship, and order. Creation naturally speaks of and points to God. Reflecting upon the nature of the created world reveals important analogies that allow the mind—illuminated by grace and revelation—to perceive the imprint of the Creator. Bonaventure (1221–74) for example declares that "these creations are shadows, echoes and pictures . . . vestiges, representations, spectacles proposed to us and signs divinely given so that we can see God."[12]

By linking the *scala naturae* with an overarching temporal-historical understanding of the world, Christians did occasionally hint in the direction of evolutionary thinking. Such hints—found most notably in the work of Augustine of Hippo (354–430)—never amounted to a consistent doctrine of biological development. Nevertheless, it is precisely the Christian sense of temporality—the feeling that history is "going somewhere"—that helped to give birth to the sense of progress that fed into sixteenth- through twentieth-century thought.

Christianity inherited its sense of temporality from its Jewish and Israelite roots. This tradition, which developed through the Old Testament and into the New, saw time not as a mere repetition of more of the same but as an ongoing movement from beginning to end, creation to eschaton. History thus becomes visible as the progressive unfolding of a divine plan. This sense of time's meaningful direction, which forms the basis of the concept of progress, contrasted sharply with the viewpoints of many pagan religions, which tended to see time as an endless repetition of timeless patterns.[13]

In his commentary on Genesis, this same sense of progress leads Augustine to suggest that creatures may have in some sense developed over time. In the first place, he recognizes that the Bible's six days of creation are highly symbolic and need not be taken too literally. In some sense, God must really have created all things simultaneously, inasmuch as God instantly established

---

12 Bonaventure, *The Soul's Journey into God; The Tree of Life; The Life of St. Francis*, trans. Ewert H. Cousins (New York: Paulist, 1978), 76.

13 See Mircea Eliade, *The Myth of the Eternal Return: Cosmos and History*, 2nd ed. (Princeton, NJ: Princeton University Press, 2005).

the rational laws of nature that govern and shape all creatures. Just as a seed has within itself the power to become an adult plant, God created plants instantaneously by imbuing the earth itself with the power of generation. This also means, however, that creatures could have taken any amount of time to develop without contradicting the six-day timeframe. Augustine writes, "In the earth, from the beginning, in what I might call the roots of time, God created what was to be in times to come."[14] In short, if God wills for there to be flowers, God's will is not frustrated if the flowers take some time to be produced. So long as God's causal influence is instantaneous and precedes time, it becomes possible to allow the passage of time to fulfill the demands of God's creative will.

This is still a far cry from the modern doctrine of evolution. In true Platonic fashion, Augustine understands the world to be perfectly governed according to transcendent forms or laws that precede the movement of time. His allowance for some temporal development flows out of this Platonic framework. It is a purposeful and nonrandom movement toward a perfect mold that was established before the creature even existed. There is no hint here that one species might become another. Just as a cherry seed is perfectly designed to become nothing other than a cherry tree—given the right circumstances—so also do all creatures unfold themselves according to a pre-given plan.

Augustine's insight was only a suggestion, and its significance does not lie in any widespread adoption by patristic or medieval readers but rather in the way it stretched out the *scala naturae* temporally. It suggested that time was always an integral factor in God's plan for biological creation. This temporal emphasis will resurface as modern ideas about biology come once again to suggest that time may be an integral factor in the formation of creatures. Moreover, Augustine's suggestion highlights three key elements that shaped medieval Christian thought and continued thus to exert influence on Western thinking even after the rise of Darwinian evolution.

First and most centrally, Christians tended to see historical developments exclusively in *teleological* terms. Built from the Greek *telos*, "end," something that is teleological is ordered toward a determinate end. That is, Christians envisioned the progress of the world not as a random chain of events, but as

---

[14] Augustine, *The Literal Meaning of Genesis*, trans. John Hammond Taylor (New York: Newman, 1982), 1:153 (bk. 5, ch. 4). See also Henry Woods, *Augustine and Evolution: A Study in the Saint's* De Genesi ad Litteram *and* De Trinitate (New York: The Universal Knowledge Foundation, 1924).

intimately tied to and oriented toward a deeper purpose at the end. Things indeed may not always remain the same, but God's providential governance of the world would never fade. World history was ultimately directed toward a perfect, future consummation of God's perfect plan.

Second, correlated to this, Christian ideas of historical development were unquestionably *anthropocentric*. Drawing on the idea of a *scala naturae*, Christians were certain that humanity held a privileged position above other creatures, and that all of creation was oriented toward human beings. There was not a sense that the history of the world's development might have turned out any other way, or that human beings might not have come into existence at all.

Such anthropocentrism actually served to validate all of creation. The *scala naturae* was meaningful and good in view of this intrinsic orientation. The ultimate goodness of the world is not found solely in its product but rather in the entire system taken together. Thus Thomas Aquinas argues in Platonic fashion that the world is more perfect because it has many different grades of creatures.[15] The world is like one perfect organism with various parts. For humanity to be the head does not degrade the proper goodness of other species. Because of this, Thomas insists that the inequality and diversity of species within the world cannot be a product of chance, but must belong to God's original intention for the world.

Hence, third, medieval Christians saw the progress of the world to be securely under the guidance and direction of a loving God. Even though Augustine suggested that there might be a kind of natural, organic mechanism to this development, any such mechanism was only really explainable in view of its Creator.

In this way, prior to the sixteenth century, it was not really possible to have a Christian "God versus science" debate. The key insight of Darwin and Wallace would not have changed this. Had they come earlier and proposed the theory of natural selection in medieval Europe, it would necessarily have been filtered through a pervasive Christian mindset that assumed from the very beginning that any natural mechanisms were already the product of a Creator God. Medieval Christianity took it for granted that nature was dynamically governed by divine personality. Any mechanical operations of the natural order fit harmoniously into this decidedly Christian metanarrative. Even the

---

15 Thomas Aquinas, *Summa contra gentiles* II, ch. 45.

Galileo incident (ca. 1610) might not have happened had Galileo published his findings a century earlier, and even during his time there were Christians who argued strongly that his heliocentric hypothesis did not contradict Scripture or tradition. The sharp opposition against Galileo was not a mere instance of medieval dogmatism but rather the slow result of significant cultural, political, and intellectual shifts that began with the sixteenth-century Protestant Reformation and led into the modern era.

Modern Systematization: Leibniz

Passing over a broad swath of history—not only in Europe but also the other 93.3 percent of the world—the most important and immediate currents underlying the science of evolution lay within modern philosophical thought. With the rise of seventeenth-century rationalism, people began to see the universe as a kind of machine, which unfolded itself on its own terms. The unifying rationality at play in Plato's thought received new vigor. In contrast to the medieval mindset, which believed that God's actions sometimes bypassed natural order,[16] Newtonian physics stressed the unwavering mathematical certainty of natural operations. Natural philosophers set out to understand how such a plurality of species might be traced back to *a priori* laws and principles from the founding of creation. Although the rise of the mechanistic worldview will eventually feed into attempts to separate theology from science, many early proponents pointed to the orderly and mechanistic operation of the universe as proof of divine providence.[17] It should come as no surprise, then, that the most influential forms of early mechanistic philosophy were still deeply spiritual; or, to put it another way, they assumed outright that the development of matter was a development of spirit.

Aristotle's ideas also found new life in the seventeenth century. Most significantly, his emphasis on the continuity of creaturely characteristics developed into the oft-quoted aphorism "nature does not make a leap" (*natura non facit saltum*).[18] John Ray brought this aphorism into modern biology as early

---

16 Aquinas, *Summa contra gentiles* III, ch. 99, §9.

17 Bowler, *Evolution*, 38–40.

18 The earliest written, formalized expression of this aphorism appears in Jacques Tissot's *Histoire veritable du géant Theutobocus, roy des Theutons* (1613), a hoax claiming the discovery of the bones of a legendary human giant, which were later identified as

as 1682 with his *Methodus plantarum nova*, and Carl Linnaeus confirmed its axiomatic status in his groundbreaking 1751 *Philosophia botanica*.[19] Darwin built upon this a century later, writing that "natural selection can only act by taking advantage of slight successive variations; she can never take a leap, but must advance by the shortest and slowest steps."[20] Interestingly, Darwin himself did not fully recognize the import of Aristotle's influence through Linnaeus until the last year of his life, when he wrote: "Linnaeus and Cuvier have been my two gods, though in very different ways, but they were mere schoolboys to old Aristotle."[21]

At the same time, the Aristotelian aphorism came to shape and be shaped by ongoing philosophical developments. Gottfried Wilhelm Leibniz (1646–1716), an influential polymath and rival to Isaac Newton, played a significant role in this regard, for he extended the aphorism into a universal "law of continuity."[22] There is a vast difference between observing that all species seem to be related by gradual similarities and declaring that *all* nature—not just biological nature—must as a rule unfold itself along a continuum. A mere observation could still admit scattered exceptions while still being effectively true. A law, on the other hand, makes a binding claim about the world and, in the case of Leibniz, about the Creator of that world. If God creates a world according to the law of continuity, then Leibniz reasons that this must be in accord with God's own goodness and perfection.

Leibniz thus helped to shape the way in which many nineteenth-century thinkers would frame the question of God's relationship to creatures. Building upon the Platonic belief that the created world must be the best and only possible image of the divine Ideas, Leibniz argued that the world cannot be anything but the most perfect. God could only have created the best of all

---

those of a prehistoric relative of the elephant. Later, in 1638, John Amos Comenius uses the aphorism as a pedagogical axiom. *De sermonis Latini studio*, ch. 16, §46.

19 John Ray, *Methodus plantarum nova* (London: Henry Faithorne and John Kersey, 1682), praef.; Carl von Linné, *Philosophia botanica* (Stockholm: G. Kiesewetter, 1751), 36 (§77).

20 Charles Darwin, *Darwin: The Indelible Stamp: The Evolution of an Idea*, ed. James D. Watson (Philadelphia: Running Press, 2005), 448. This anthology contains Darwin's most important works, including *The Origin of Species* and *The Descent of Man*.

21 Letter 13697 to William Ogle, February 22, 1882, in *Life and Letters*, 3:252.

22 Bentley Glass, "The Germination of the Idea of Biological Species," in *Forerunners of Darwin, 1745–1859*, ed. Bentley Glass, Owsei Temkin, and William L. Straus (Baltimore: Johns Hopkins Press, 1968), 38–39.

possible worlds precisely because God's will, ordered according to the "principle of the best," always acts in accord with the highest reason.[23] Leibniz's world was thus a kind of grand machine; not only did the world as a whole flow out from God in an organized, mechanistic way, but so also did individual organisms develop mechanically, without additional supernatural interference, from the biological seed.[24] The ongoing operation of this cosmic machine beautifully and dynamically maintained itself in physical, *meta*physical, and moral perfection.[25]

Leibniz's intention was not merely to justify the present state of the world but also to justify the moral perfection of God as its author. He coined the now common term "theodicy." *Theodicy* refers to the question about or an explanation of how we reconcile the existence of evil with the goodness of God. For Leibniz, evil, suffering, and violence would necessarily contradict the notion of a just and loving God unless such evils can be incorporated back into God's own perfection as the natural and even mechanical consequence of God's perfect plan for the world.

Building on this framework, Leibniz made at least two important advances toward a Darwinian idea of evolution. First, he framed the question of the world in view of theodicy. This emphasis, which Darwin will inherit, changed the demands of a suitable cosmology in a way that inherently put God on trial. Like Darwin, Leibniz had no interest in abandoning the notion of God. Rather, he felt that the mechanical nature of the world sufficiently insulated God from being directly implicated in evil. To the extent to which the world is a machine that runs independently of God's direct intervention, God could not be held responsible for everything that the world entails.

Second, unlike the Neoplatonists, Leibniz did not characterize complexity as a fall away from simplicity. In fact, there is a sense in which the perfect simplicity of God *necessitates* the development of complex and interrelated systems. Perfect simplicity was for Leibniz not the absence of differentiation but rather the ability to incorporate and represent diversity within itself without internal division. This fed into his understanding of the world's mechanical development as an overall positive movement.

---

23 Gottfried Wilhelm Leibniz, *Theodicy: Essays on the Goodness of God, the Freedom of Man, and the Origin of Evil*, ed. Austin Farrer (London: Routledge, 1952), 61, 68, 128; *Leibniz's Monadology: A New Translation and Guide*, trans. Lloyd Strickland (Edinburgh: Edinburgh University Press, 2014), 23.

24 Leibniz, *Theodicy*, 66.

25 Leibniz, *New Essays*, 696.

Nevertheless, Leibniz's views still carried two major limitations that would continue to shape the way that people thought about worldly development. First, like Plato, Leibniz's concept of history was relatively limited. The transition from simple to complex was more a matter of the transition from Ideal to real than of past to present. Second, Leibniz's view was still inherently teleological. At all times, the process of complexification was governed from the beginning by the principle of the best, which necessitated that the world that it produced must be the best of all possible worlds. Any biological development thus had to be explainable according to this same paradigm. Certainly not every individual organism had to be perfect, but ultimately they had to all fit into a perfect whole that expressed the *a priori* unity and perfection of its Creator.

The Transition from Philosophy to Biology

This brief history has highlighted a vast array of hints from philosophy and Christianity that pointed Western thought toward viewing the world in terms of historical progress. Nevertheless, the critical transition toward applying such a progress-centered framework as an interpretive lens for the specific, empirical data of biology did not occur all at once. From the standpoint of intellectual history, this transition occurred in a subtle manner precisely because the boundary between philosophy and what was once called "natural philosophy" was thin. It was only natural and traditional for naturalists to interpret biological observations in terms of broader philosophical and metaphysical commitments that had long since refracted the ways in which modern people understood the natural world. Yet, as we have seen, Western thought was ambivalent inasmuch as it contained the roots both of seeing species in typological terms as static, unchanging, and therefore ahistorical forms and of seeing the world as a living organism progressing toward some higher self-realization through the various species that are, by analogy, the members or organs of this cosmic body.

The perspective of Darwin and Wallace thus came to be posited and accepted not so much despite Western thought but rather because of a shift within it from one pole to another. This shift cannot be precisely timed, since it did not happen all at once. Yet that there was a shift is evidenced by a popular surge in evolutionary systems of thought well prior to the theory of natural selection. Wallace and Darwin published their insights within a culture and intellectual climate that was already becoming obsessed with progress and development. While to some extent this thinking provided for a more positive reception of natural selection, it also all but guaranteed that the scientific theory would be misunderstood and misappropriated. Some thinkers would point to the facts of natural selection as evidence for their own philosophical and political forms of evolutionism. In this way, Darwin's theory became a convenient empirical justification for others' deep-seated ideological convictions.

THE MEANING AND RECEPTION OF DARWIN'S IDEA

So why does this nonscientific prehistory of Darwinian evolution matter? Was it merely a passing influence, which provided Darwin with some inspiring thoughts that ultimately led him to his discovery? On the contrary, even though the ideas of Darwin and Wallace were shaped at least indirectly by this philosophical background, much more important is the way in which it formed the minds and hearts of those who received their discovery. To see this, first we have to acquire a sense of what they taught and why it was so novel. Nevertheless, given the limits of space and specialization, we cannot hope to provide a complete and detailed scientific account of evolution, nor can we catalog the wealth of important developments in evolutionary science up to the present day. For that, the reader must turn to any of the many expert scientific books dedicated to the topic. For our purposes here, a brief and cursory sketch of key *thematic* points must suffice for exploring the specifically theological significance of evolution. Moreover, we will focus on Darwin rather than Wallace because of the former's popular significance.

Darwin's Natural Selection

In an irony that surprises many, Darwin routinely avoided the term "evolution." This was at least in part because of the way in which the word was already entrenched in both popular use and technical literature.[26] "Evolution"

---

26 Stephen Jay Gould, *Ever Since Darwin: Reflections in Natural History* (New York: Norton, 1977), 34–37; see Bowler, *Evolution*, 8.

was established by the end of the eighteenth century—before Darwin's time—as the preferred term for different kinds of progressive development, especially through gradual changes. Derived from the Latin *ēvolvere*, "to unfold" or "roll out," the word in its technical usage implied a design-driven linear progression. It was used thus to describe viewpoints that were actually incompatible with Darwin's own theory.

The key difference here is *progress*. While Darwin ultimately provided the key scientific basis for understanding gradual biological changes, his understanding of such changes did not necessarily imply progress or any kind of linear transition from worse to better.[27] He instead utilized terms such as "descent with modification" and "transmutation" because they lacked a sense of direction or higher purpose. In this sense, Darwin's key contribution was not a theory of "evolution" as such but rather the theory of *natural selection*: an understanding of how species can change through a purely natural and chance-driven process.

Darwin's insight depended upon his ability to envision the buildup of minute biological mutations, over a multitude of generations, into dramatic adaptations on a much larger scale. This is akin to imagining a tiny drip of water creating a sizable stalactite over a hundred thousand years, except that with large animals, every "drop" would take generations and the entire process must be stretched out over hundreds of millions of years.

That creatures change in small ways from generation to generation was obvious, but compiling these changes into a workable process of "descent with modification" posed at least two conceptual, scientific problems. First, Darwin had to show how apparently random mutations could tend in any one specific direction. A foal may grow to be larger than the mare, but that of itself does not explain how the modern-day horse could have descended from something the size of a small dog. Thomas Malthus, one of Darwin's key influences, acknowledged in 1798 that one may breed a sheep to have a smaller head and legs, but he insisted that there must be a limit; "were the breeding to continue for ever, the head and legs of these sheep would never be so small as the head and legs of a rat."[28] Second, Darwin had to allow an enormous and at the time incomprehensible span of time for such selection to arrive at the present day's vast diversity of species. Contrary to popular

---

27 On the complexity of Darwin's sense of progress, see Bowler, *Evolution*, 146.

28 Thomas Robert Malthus, *First Essay on Population, 1798*, ed. James Bonar (London: Macmillan, 1926), 164–65.

history, Christians in the nineteenth century were not widely invested in James Ussher's famous estimate that the world began in 4004 BCE. Nevertheless, Darwin's suggestion that the geology of a certain region of southeast England evidenced "a far longer period than 300 million years" still exceeded the wildest dreams of leading nineteenth-century geologists.[29]

Darwin's theory thus contradicted many of the leading scientific views of the time. His was an uphill battle. In order to even imagine the process of natural selection, Darwin had to proceed by analogy from the limited experiences of the time to imagine a purely natural process that would explain and transcend these experiences. To say that this was a feat of imagination is not to downplay the empirical character of Darwin's research. Rather it was precisely Darwin's empirical findings that allowed him to posit something that flew in the face of what many considered to be mere common sense. By observing dramatic changes that must have come about through mere gradual mutations, Darwin was able to describe a process of natural selection that lacked any requirement for an intelligent, free agent making the selection.

By comparison, humans have been practicing *artificial selection* for millennia. When an edible plant produces offspring, a keen agriculturalist will select the ones that best suit our purposes: the best taste, the best color, the best yield, and so forth. Those that are less desirable are plucked out and prevented from creating offspring. Over many generations, by eliminating nonselected seedlings and allowing only the selected plants to thrive, we can produce a dramatically different kind of plant. Evidence has shown, for example, that humans' artificial selection helped maize to develop from a wild grass that yielded a few small kernels in a hard case into a vibrant crop that yields a large, exposed cob with a mass of sugary kernels.[30] Since artificial selection has modified maize according to human needs rather than those of the plant itself, it has also left maize with relatively poor qualifications for surviving in the wild. Modern maize largely depends upon humankind for its propagation.

What artificial selection accomplishes in a small way, natural selection does on a much larger scale and over longer periods of time. Although the

---

29 Darwin, *Indelible Stamp*, 495. See also Joe D. Burchfield, "The Age of the Earth and the Invention of Geological Time," *Lyell: The Past Is the Key to the Present (Geological Society, London, Special Publications)* 143, no. 1 (1998): 140.

30 Chin Jian Yang, Luis Fernando Samayoa, Peter J. Bradbury, et al., "The Genetic Architecture of Teosinte Catalyzed and Constrained Maize Domestication," *Proceedings of the National Academy of Sciences* 116, no. 12 (March 19, 2019): 5643.

science of DNA arrived long after Darwin's time, he understood that somehow procreation resulted in small changes from generation to generation. Darwin argued that beneficial mutations tend to be preserved and emphasized accidentally by the demands of an organism's environment. Organisms tend to produce more offspring than the environment can support, which leads to a "struggle for existence."[31] Individual organisms must contend with other organisms, other species, and even with the raw conditions of life. Many—if not most—of an organism's offspring will die. Which ones die and which survive to multiply will be decided at least in part by the relative adequacy of biological mutations. If some of the offspring are less well-adapted to their environment, then they stand a higher chance of dying. Conversely, mutations that happen to be beneficial for the given circumstances, by helping to preserve the single organism, will stand a greater chance of being spread to its descendants.

For example, imagine that a flowering plant species grows in an area that, because of changes in weather, is becoming more prone to frost. Some of the individual plants are able to resist low temperatures while others are not. If these poorly adapted plants die from frost before they are able to reproduce, then the next year's garden will have more of the frost-resistant variety. Over time, if the frost continues, the frost-resistant strain may become dominant, and the other strain may become extinct. In sum, natural selection means that, inasmuch as environmental factors work against other strains, the more well-adapted organisms are as it were "selected" for propagation by the environment according to their biological traits. Small changes like this, over millions of years, can result in dramatic biological diversity.

Darwin's key insight, then, was the realization that such a simple and mechanical law could be the basis for all of the different species of plants, animals, and fungi that we observe in the world. Natural selection made it possible to conceptualize such changes by showing how they are directed by environmental factors. That is, without there being any overall purpose and general direction evident in biological change, there is the very basic and observable local direction of particular traits. The process by which particular traits become dominant within a population is directed toward nothing other than the immediate survival and prosperity of that population.[32]

---

31 Darwin, *Indelible Stamp*, 380.
32 To be clear, Darwin's far-reaching reliance on natural selection will not altogether survive in later evolutionary science. There can for example be cases where a species

Teleology and Darwin's Theodicy

Darwin did not propose his theory of natural selection as a metanarrative. A metanarrative is by definition explanatory, and in that sense it is teleological. A teleological explanation sees things as ordered toward a definitive end (Greek *telos*), which makes the process of achieving this end something purposive and meaningful.

In contrast, Darwin's theory on the whole does not point in any one definite direction. It is *dys*teleological.[33] The ongoing modification of a species is driven by nothing more than the immediate and concrete demands of a particular environment. It is not ordered toward some ultimate, perfect end, but only toward the more basic and immediate goal of perpetuating the survival of a group of organisms within its environment. Because of this, natural selection cannot produce a perfect species, but only one that is relatively better adapted to a situation than another. Darwinian natural selection is random then only in the specific sense that does not unfold in any predetermined direction or according to an *a priori* divine plan. Rather, as Peter J. Bowler explains, it is "a process of trial and error based on massive wastage and the death of vast numbers of unfit creatures."[34]

This is not to say that natural selection has no observable patterns or tendencies, but only that it is not governed by some overarching tendency toward a grand, divine plan. Darwin himself was not altogether opposed to God or the idea of a divine plan,[35] but he felt that the problem of natural evil deserved an explanation independent of divine intervention. This perspective allowed the mechanism of natural selection to be wholly understandable from the perspective of the natural sciences without the presupposition of any specific divine governance. Or to put it more starkly: not only does Darwin's schema not require God, it really has no place for God within it. There is no gap or seam where one might suggest, "Maybe that is where God intervenes."

Following already established trends, Darwin tended toward a dry separation of science and religion into distinct, autonomous spheres. For him,

---

survives and propagates not so much because it is well-adapted but merely because it is not *mal*adapted.

33 See Mary Midgley, *Evolution as a Religion: Strange Hopes and Stranger Fears*, rev. ed. (London: Routledge, 2002), 38. As Alister McGrath points out, Darwin did not reject all teleology but came to a broader understanding of teleology that points to the intermediate end of adaptation. Alister E. McGrath, *Darwinism and the Divine: Evolutionary Thought and Natural Theology* (Malden, MA: Wiley-Blackwell, 2011), 162–63.

34 Bowler, *Evolution*, 6.

35 McGrath, *Darwinism*, 160.

science occupied itself with empirical data, and for that reason it could not directly bear upon theological questions. "My opinion," he wrote, "is not worth more than that of any other man who has thought on such subjects."[36]

Despite his reservation, Darwin could not altogether avoid making a theological statement. As John Haught points out, to the extent to which Darwin proposed natural selection as an alternative to God's direct intervention—as a different answer to the same basic question of origins—his science remained bound to theology.[37] After all, to insist that God does *not* do a certain thing is still to make a theological statement. There is a fine line between claiming that we need not assume God's involvement and insisting that God simply cannot be involved at all; it is hardly surprising that Darwin did not remain altogether to one side. From the point of view of early religious critics, Darwin's God-free approach represented not simply a theological trespass but also a scientific error. Catholic critic Richard Simpson argued in 1860 that Darwin improperly lines up a wealth of empirical facts as would-be proof for the all-too-metaphysical claim that all organisms necessary derive from one primeval original.[38] Simpson's critique was not limited to Darwin, however, but also targeted the earlier views of Jean-Baptiste Lamarck (1744–1829) and Robert Chambers (1802–1871).

For Darwin, removing any reliance on God for the operation of biological changes was necessary not simply because such divine intervention could not be empirically demonstrated but also because this approach allowed him to face a particularly daunting theological question: the problem of evil. How could the benevolent governance of God be directly involved in a world where animals cruelly prey upon one another? "I had no intention to write atheistically," he wrote, but the gross suffering caused by predators and parasites made it hard for him to see the natural world as directly planned by a loving God:

> There seems to me too much misery in the world. I cannot persuade myself that a beneficent & omnipotent God would have designedly created the Ichneumonidæ with the express intention of their feeding within the living bodies of caterpillars, or that a cat should play with mice.[39]

---

36 Letter 5307 to M.E. Boole, December 14, 1866, in Darwin, *Life and Letters*, 3:67.
37 Haught, *Making Sense*, 17.
38 Richard Simpson, "Darwin on the Origin of Species," *The Rambler*, March 1860, 363–65; see also John Lyon, "Immediate Reactions to Darwin: The English Catholic Press' First Reviews of the 'Origin of the Species,'" *Church History* 41, no. 1 (1972): 78–93.
39 Letter 2814 to Asa Gray, May 22, 1860, in *Life and Letters*, 2:312.

Removing God from biology not only allowed for a consistently empirical approach; it allowed Darwin to see the chaotic horrors of death, pain, and suffering in the natural world as oriented not toward some grand divine plan but toward the more immediate goal of adaptations and survival. Such an approach makes no promise that these evils will ultimately result in a higher good, but it does at least insulate God from responsibility. If God did not directly will for the parasite to torture its host, then this situation is nothing more than a happenstance of natural adaptation. One could always suppose—as Darwin apparently did—that God was still in an ultimate way the original cause of the universe as a whole.[40] For Darwin, however, natural selection made it so that the untidy particulars of that universe no longer posed a challenge to acknowledging the basic goodness of its original cause.

Spencer: Survival of the Fittest

If Darwin seldom used the term "evolution," then how did we end up with a Darwinian theory of evolution? Or to put it better, how did an essentially dysteleological view of biological change end up promoting a teleological metanarrative of evolution?

To begin, even Darwin himself was never quite perfect in his dysteleological convictions. Unable to fully disown the hierarchical scale of organisms inherited from Platonism, Christianity, Leibniz, and even the earlier evolutionary views of Lamarck, Darwin still spoke about humanity as occupying "the very summit of the organic scale."[41] This assumption was tied into his particular social, economic, and cultural context. Standard Victorian racism made it easy for Darwin to think of the "refined" Englishman as more perfectly evolved than the "savages" outside of Europe. While Darwin never pretended that natural selection offered any easy answers for the further improvement of the human race, he did believe that only vigorous competition would allow "more gifted men" to become "more successful in the battle of life." In short, Darwin never fully abandoned deeply engrained cultural

---

40 Peters and Hewlett, *Theological and Scientific*, 58.
41 Darwin, *Indelible Stamp*, 1053–54. See also Francisco J. Ayala, *Darwin's Gift to Science and Religion* (Washington, DC: Joseph Henry, 2007), 34; David N. Livingstone, *Darwin's Forgotten Defenders: The Encounter Between Evangelical Theology and Evolutionary Thought* (Grand Rapids, MI: Eerdmans, 1987), 48.

prejudices about the hierarchy of creatures, and because of this he could not altogether eliminate the teleology of natural selection.

Perhaps this is why Darwin also failed to untangle the perception of his views from his early influential readers, who shaped the way in which evolution entered into cultural consciousness. These include Ernst Haeckel (1834–1919) and Herbert Spencer (1820–1903) among the German- and English-speaking worlds, respectively.[42] While we cannot here examine in full detail the reception of Darwin's thought by early interpreters, it is important to see how such interpretations immediately expanded Darwin's claims beyond the confines of strict, empirical science. What Darwin actually *said* was not necessarily what his more fervent supporters *heard*.

Herbert Spencer had already laid out his own theory of evolution on a predominantly philosophical basis before the publication of Darwin's groundbreaking work. For Spencer, "evolution" referred to a universal dynamism of development and adaptation that served not merely as an empirical qualification of biological change, but as an all-encompassing philosophical principle. This principle unified science and religion into a complete philosophical system that connected physics, biology, psychology, sociology, and ethics.[43] Importantly, Spencer "did not borrow the idea of evolution from biology in order to extend it to the universe. He started with the universal evolution, in which he afterwards included biological evolution."[44]

For Spencer, reality is structured by "equilibration," an internal and universal drive toward establishing and maintaining equilibrium or balance. Just as there is a direct relationship between the amount of fuel an engine consumes and the work it produces, so also there must be a balance between the biological organism and the coincidental factors of its environment that impel it to stay or adapt. If its environment changes, so also must the organism change to adapt. Borrowing a mathematical term, Spencer argues that such adaptation represents the dynamism of "moving equilibrium," a kind of homeostatic interplay of forces that, on the whole, ensures that development or change is the ultimate constant.[45]

---

42 See Gould, *Ever Since Darwin*, 36–37; Richard Weikart, "The Origins of Social Darwinism in Germany, 1859–1895," *Journal of the History of Ideas* 54, no. 3 (1993): 475–76.

43 Herbert Spencer, *First Principles* (New York: De Witt Revolving Fund, 1958), 34–35.

44 Ernest Barker, "The Scientific School: Herbert Spencer *and* After Spencer," in *Herbert Spencer: Critical Assessments*, ed. John Offer, vol. 4 (New York: Routledge, 2000), 7.

45 Spencer, *First Principles*, 480–84.

In view of this, Spencer received Darwin's groundbreaking theory enthusiastically. Promoting Darwin's work as the empirical confirmation of his own system, he painted natural selection as equivalent to what he termed "indirect equilibration," a balancing that occurs negatively when environmental factors lead to the extinction of organisms that fail to properly adapt to changing situations. This led Spencer to coin the phrase "survival of the fittest," which he offered as a synonym for Darwin's natural selection.[46] Wallace actually encouraged Darwin to adopt Spencer's phrase, for he felt that people tended to carry the analogy of selection too far; they took it as implying that some higher intelligence (i.e., God) does the selecting.[47] Of course, the word "fittest" presented its own difficulties; natural selection does not produce plants and animals that are most fit on an absolute scale, but only relatively better adapted to survive to their present environment. Regardless, Darwin adopted Spencer's phrase in 1869.[48]

Language has a way of getting away from us, however, and Spencer's phrase illustrates this well. Not only did "survival of the fittest" fail to provide the kind of safeguard that Wallace had intended, but it also encouraged other misinterpretations through association with Spencer's philosophy. Spencer's system is built upon conflict.[49] It sees existence as the ongoing and progressive struggle of fundamental opposites. Spencer's "moving equilibrium" is the synthesis of matter's primordial antagonistic forces of attraction and repulsion, cohesion and resistance, order and chaos.[50] "Survival of the fittest" thus serves to highlight the basic *agonism* or competitiveness of Spencer's system: Struggle is internal to order; violence is necessary for peace. Spencer's phrase thus not only promoted the false impression that Darwinian evolution was steeped in a sense of progress; it also positioned the Darwinian paradigm to become an all-explanatory and even prescriptive principle of meaning.

---

46 Herbert Spencer, *The Principles of Biology* (London: Williams and Norgate, 1864), 1:443–45, 474.

47 Letter 5140 (190) from A.R. Wallace, July 2, 1886, in *More Letters of Charles Darwin: A Record of His Work in a Series of Hitherto Unpublished Letters*, ed. Francis Darwin and A.C. Seward (London: Murray, 1903), 1:267–70.

Letter 5145 (191) to A.R. Wallace, July 5, 1886, in *More Letters*, 1:270–71; also *Life and Letters*, 3:45–47.

48 Charles Darwin, *On the Origin of Species by Means of Natural Selection, or the Preservation of Favoured Races in the Struggle for Life*, 5th ed. (London: Murray, 1869), 72.

49 In this way, it is reminiscent of Empedocles. Barker, "Scientific School," 7, 15. See note 4 above.

50 Spencer, *First Principles*, 229. For more on dialectic, see chapter 3.

Chaos makes sense in view of order; violence becomes meaningful in light of peace; death is of fundamental necessity for life.

EVOLUTION AS METANARRATIVE

Because of this, as word of Darwin's discoveries spread, what many people heard was quite the opposite of what he taught. They took it to mean that species *make progress* by means of selection. The teleological context overrode Darwin's essentially dysteleological message. Humanity was held to be a "more evolved" sort of animal, higher on the ladder of progress, further along on the conveyor belt of the evolutionary machine.

For this reason, Stephen Jay Gould argues that Darwin's theory of natural selection really failed to take hold in his own lifetime. Its denial of progress made it unpalatable to a public that expected organic development to be understandable in such terms. Natural selection did not meet public demand because it "proposes no perfecting principles, no guarantee of general improvement; in short, no reason for general approbation in a political climate favoring innate progress in nature."[51] Even Darwin himself came to soften his stance. While he originally treated natural selection as the primary means of biological change almost to the point of exclusivity, he later allowed other factors (such as sexual selection and the use or disuse of organs) a greater share in the credit.[52]

Thus, contrary to Darwin's original intention, these factors led people to see evolution by natural selection as a meaningful, teleological process. Not only this, but it also made selection into the very standard of meaning. Many saw natural selection as choosing creatures that are *better* than others—not merely better adapted for the particular circumstances, but actually in some sense better on an objective scale. In short, Darwin's theory of evolution

---

51 Gould, *Ever Since Darwin*, 45.
52 Charles Darwin, *The Origin of Species by Means of Natural Selection, or the Preservation of Favoured Races in the Struggle for Life*, 6th ed. (London: Murray, 1872), 421. Also, whereas in the first edition he averred that "Natural Selection has been the main but not exclusive means of modification," he subtly changed this to "the most important, but not the exclusive, means of modification" (6). St. George Jackson Mivart criticizes this inconsistency as undermining Darwin's frequent dogmatic claims about humanity. *Essays and Criticisms* (London: J.R. Osgood, 1892), 1–59.

became, as quickly as it was received, yet another evolutionary metanarrative. It became evolutionism, the narrative to end all narratives, which purported to govern the meaning and value of all existence.

It is no surprise then how quickly this specifically biological idea flooded into every aspect of human thought. History, language, music, culture, religion, and even the Bible came to be seen as products of evolutionary developments. The evolutionary viewpoint unlocked new and important ways of viewing these disciplines.

Because of this, because evolution was received as a metanarrative that spoke to the meaning of humanity and the world, it tended from the very beginning to be accepted or rejected *on these terms*. For most people, evolution did not stand upon the validity of a set of empirical facts. If it was true, it was true because it was meaningful; if false, it was false because it jeopardized the meaning of the world (as represented in the Bible, for example).

The Human Dimension

Just as today, many of the early reactions against Darwinism depended upon religious convictions. Nevertheless, the situation was never so simple as popular histories tend to suggest. In fact, Darwin's principal opponents in his day were not religious fanatics but members of the scientific establishment, while many key supporters were religious or even clergymen.[53] Ironically, although Darwin's discovery spurred many of his contemporaries to accept the claim that species change and develop, they often continued to reject Darwin's particular theory of natural selection on both scientific and theological grounds.[54]

---

53 Some examples include James Dwight Dana, Asa Gray, Aubrey Moore, and George Frederick Wright. See Livingstone, *Darwin's Forgotten Defenders*, 60–77; Midgley, *Evolution as a Religion*, 12; Stanley M. Guralnick, "Geology and Religion Before Darwin: The Case of Edward Hitchcock, Theologian and Geologist (179–864)," *Isis* 63, no. 4 (1972): 542–43; James R. Moore, *The Post-Darwinian Controversies: A Study of the Protestant Struggle to Come to Terms with Darwin in Great Britain and America, 1870–1900* (Cambridge: Cambridge University Press, 1979). See also A. Hunter Dupree, "Christianity and the Scientific Community in the Age of Darwin," in *God and Nature: Historical Essays on the Encounter Between Christianity and Science*, ed. David C. Lindberg and Ronald L. Numbers (Berkeley: University of California Press, 1986), 351–68.

54 Chauncey Wright, *Darwinism: Being an Examination of Mr. St. George Mivart's "Genesis of Species"* (London: John Murray, 1871), 4–5.

Whether yea or nay, religious responses to natural selection came to center upon its perceived relevance with regard to the meaning and dignity of the human person. As William Jennings Bryan, a famous opponent of evolution, once said, "The only part of evolution in which any considerable interest is felt is evolution applied to man. A hypothesis in regard to the rocks and plant life does not affect the philosophy upon which one's life is built."[55]

Darwin was at first reticent to state the obvious conclusion that if natural selection applies to other organisms, it should logically apply to humans too. He knew that this would open up dangerous avenues of criticism. When he did eventually discuss human evolution, he was careful to avoid touchy theological questions such as the origin of the human soul. If we are unable to pinpoint the precise moment when a proto-ape becomes a human in the full, spiritual sense, he argues, this is no more problematic than science's inability to delineate at what moment a human embryo acquires an immortal soul.[56]

Yet Darwin's early reticence in regard to human evolution is ironic, since his insight in regard to animal evolution was facilitated by his reading of Thomas Malthus's influential views on population growth and competition among *humans*, which he first published in his 1798 *Essay on the Principle of Population*.[57] Darwin accepted Malthus's views on the tendency for human populations to grow exponentially and applied this same logic to plants and animals.[58] If populations tend to overproduce offspring, then it becomes possible to imagine that selection could occur through a corresponding high rate of mortality.[59] When the population tends to grow beyond the available means of sustenance, leading to competition over resources, to suffering, and to death, this tendency could be seen as privileging or "selecting" those individuals who are better suited for survival in the given conditions. Malthus thus opened Darwin's eyes to seeing individuals within a population

---

55 William Jennings Bryan, "God and Evolution," *New York Times*, February 26, 1922, 84.

56 Darwin, *Indelible Stamp*, 1050.

57 Sandra Herbert, "Darwin, Malthus, and Selection," *Journal of the History of Biology* 4, no. 1 (1971): 217. Some scholars have downplayed the role of Malthus's ideas, but Herbert illustrates from Darwin's journals how reading Malthus proved to be a turning point in Darwin's intellectual development.

It should be noted that Malthus's first edition (1798) of *Essay* is substantially different from all later editions, effectively comprising a different work despite the shared name. Darwin owned a copy of the sixth edition.

58 Darwin, *Indelible Stamp*, 348; *Autobiography*, 120–21.

59 Herbert, "Darwin, Malthus," 212.

as being fundamentally in competition. Darwin paid homage to Malthus's influence by adopting his key description of life as "a struggle for existence."[60] By doing so, he helped to enshrine the concept of competition as a fundament of biological evolution.[61]

As Darwin and other thinkers came more and more to see concepts pertaining to human populations as applicable to animals, and vice versa, this served to downgrade the difference between humans and animals from a difference in kind to a difference in degree. Since that time, many have hailed Darwin as the forerunner who, by hammering a critical fault into the lofty edifice of human vanity, set in motion the eventual destruction of the old, anthropocentric point of view. Having once enjoyed a privileged place atop the great chain of being, humanity now stood in danger of being seen as the incidental product of natural processes and conditions. Francisco Ayala thus hails Darwin's discovery as a continuation of the Copernican Revolution: just as Nicolaus Copernicus removed the earth from its privileged place among the stars, so also Darwin demoted humanity to an insignificant role within the greater drama of evolution.[62] Darwin thus became a symbol of a new way of conceptualizing humanity's role within the universe.

In fact, even by the end of the nineteenth century, the widespread idea popularly labeled "Darwinism" was not the teaching of natural selection as such, but the idea that humans are essentially nothing more than animals who happen, through accidents of selection, to have developed a superior intelligence. Vladimir Solovyov highlights what "Darwinism" meant to the average Russian with a joke. An energetic and self-assured "Darwinist" takes it upon himself to educate an ordinary merchant from Moscow. After ranting extensively about humanity's basic equality with the animals, the intellectual notices the merchant's blank stare. Annoyed at the merchant's silence, he asks:

— Understood?
— Understood.
— Well, what do you have to say?
— What's left to say? Why, if I'm a dog and, well, you're a dog too, then what kind of a conversation can a dog have with a dog?[63]

---

60 Malthus, *First Essay on Population, 1798*, 47–48.
61 Bowler, *Evolution*, 104–105, 161–62.
62 Ayala, *Darwin's Gift*, 40–41.
63 Vladimir Solovyov, "The Idea of a Superman," in *Politics, Law, and Morality*, trans. Vladimir Wozniuk (New Haven, CT: Yale University Press, 2000), 263.

Many opponents of evolution have focused excessively on how the timeline of evolution disagrees with the literal chronology of events in the book of Genesis. In fact, the more serious challenge that evolution presents to the Bible regards Genesis's starkly anthropocentric perspective. Genesis does not provide one chronology of creation but two, and although the two disagree about the order of events, both are structured so as to emphasize humanity's pride of place as the jewel of all creation. In the first account God saves the best for last, creating man and woman in God's very own image (Gen 1:24–31). In the second account, God creates man prior to the animals so that he can have the honor of naming the various animals as they are created. This time, the best that God saves for last is woman. She is fashioned last because she alone represents a suitable companion for man (Gen 2:18–25).

This anthropocentric principle was not only central to traditional Christianity, but it also in many ways served as the centerpiece of early modern thought and culture. Much of modern philosophy was marked by a resolute "turn to the self," an engagement with reality from the perspective of the thinking subject. Accordingly, it only made sense to read all of reality as ordered toward its domination by and perfection in rational inquiry on the part of the (human) subject. For this reason, in Darwin's day it was common even for leading evolutionists to make an exception of the human race. Alfred R. Wallace, for example, argued that natural selection held true for everything except the human brain.[64]

Accordingly, despite the overwrought stereotype that sees evolution as a battle between science and religion, the more consequential divide was between those who saw evolution as downgrading humanity from its place in the clouds and those who saw it as maintaining or supporting this place. Religion and philosophy naturally played into both parties.

In this way, the human dimension of evolution became the nexus of the conflict over the idea of evolution. For different parties and in different ways, humanity came to signify the place where the question of meaning comes to the fore.

This question of meaning more than ever took on a temporal dimension. The traditional Christian teleological perspective saw humanity as meaningful both *a priori* and *a posteriori*—both in its origin and in its ultimate end. The idea that humanity might be the unintended result of a chaotic, meandering, and impersonal mechanism of natural selection put *a priori* meaning into question, which threatened to undermine *a posteriori* meaning

---

[64] Gould, *Ever Since Darwin*, 50; Darwin, *Indelible Stamp*, 646.

as well. If humanity was never planned nor intended, what significance could the race—or even a single man—have within the grander scheme of time?

Alfred Lord Tennyson struggled with this prior to 1849. Far from the caring mother, nature appears indifferent and even hostile to species and individuals. He writes:

> Are God and Nature then at strife,
> That Nature lends such evil dreams?
> . . .
>
> 'So careful of the type?' but no. . . .
> She cries, 'A thousand types are gone:
> I care for nothing, all shall go.'
>
> . . . And he, shall he,
> Man, her last work, who seem'd so fair, . . .
>
> Who trusted God was love indeed
> And love Creation's final law—
> Tho' Nature, red in tooth and claw
> With ravine, shriek'd against his creed—
>
> Who loved, who suffer'd countless ills,
> Who battled for the True, the Just,
> Be blown about the desert dust,
> Or seal'd within the iron hills?[65]

The meaning of humanity thus had everything to do with its future: Will humans survive? Or can a world that once existed without us still continue to spin after we are gone?

Evolution, Violence, and Meaning: Social Darwinism

As we shall see in later chapters, many diverse strategies exist today for dealing with this problem of meaning. For now, it is important to see that the way to

---

65 *In Memoriam A.H.H.*, LV–LVI.

such strategies was paved by the reception of Darwin by various advocates of progress, who dealt with the abyss of *a priori* meaning by looking optimistically into the future. Spencer, for example, sees the imperfection of the human race as an indication that the process of evolution is as yet incomplete. As we continue to adjust to the demands of civilized society, traits like physical strength and agility are bound to become less important and to fade behind the ongoing development of humanity's mental and moral acuity. In a bout of self-assured prophecy steeped in Victorian racism, Spencer argues that less-refined humans, failing to adapt to the requirements of modern society, will tend to over-reproduce and stretch limited resources until they bring about their own extinction. Their death will ensure for superior (i.e., aristocratic English) humans "a constant progress towards a higher degree of skill, intelligence, and self-regulation . . . a more complete life."[66]

In effect then, Spencer's response inserts evolution itself into the void of meaning left empty by the decline of *a priori* reasoning. To believe that evolution must be trending toward something better, something higher, allows Spencer to justify the messy and chaotic process by which this trend is prosecuted. This in turn generates a new kind of *a priori* meaning, namely, the very principle of evolution itself, distilled into the survival of the fittest. Humans are meaningful to the extent to which they preeminently embody such fitness through mutual competition.

Ideas like this highlight the violent aspect of natural selection. If nature is defined by the survival of the fittest, then there must be a sense in which proving one's own fitness through competition is proper to the natural order.[67] Although Spencer, like Darwin, believed that morality was due to progress in an evolutionary way, this naivete made little sense in view of how Spencer's ideas enshrined violence as the arbiter of meaning. It was therefore inevitable that Spencer's words would promote the development of *social Darwinism*.

The term *social Darwinism* has signified many different things since its inception. Immediately following Darwin's work, many different ideologies were interested in utilizing the growing popularity of evolution as a justification for their own political and social views. Wildly different views, ranging from socialism to conservatism, sought to claim Darwin as their own and in that sense could be called varieties of social Darwinism. Today, however,

---

66 Spencer, *Principles of Biology*, 2:495–500.
67 Midgley, *Evolution as a Religion*, 7.

"social Darwinism" typically refers more specifically to any ideology that uses the competitive aspect of natural selection to validate social and economic struggle. Such a view argues that the struggle of society leads to a gradual improvement of the race through the elevation of worthier and better adapted individuals over poorer, inferior human beings.

Understood in this sense, social Darwinism did not arise as a unified and cohesive movement. Its ideological foundations—the social and economic glorification of competition—were laid long before Darwin's groundbreaking theory came onto the scene. As natural selection began to occupy the cultural imagination, many people sought to take advantage of this situation by allying their own particular ideology with the science of natural selection. Social Darwinism as we know it was only one such attempt, but it made an important impression. Its advocates easily found their views echoed in the competitive language of Darwin and Spencer. However, the idea of social Darwinism was also built up by its critics, who either saw this as a misappropriation of natural selection, or else as further proof of a basic and pernicious materialism underneath Darwin's theory.[68]

Spencer did not take social Darwinism to its utmost extremes, nor did he use the term himself. In fact, many early readers of Spencer saw his views as promoting altruistic rather than competitive behavior.[69] Nevertheless, he contributed in his small way to the gradual diffusion of the idea that natural selection justifies and even prescribes competitive behavior within society as necessary for the healthy development of the human race.[70] The survival of the fittest naturally advances those humans whose drive for self-preservation is the strongest and most developed, allowing greater correspondence between their abilities and the demands of civilized life.

As early as 1862, Clémence Royer, in her preface to the first and only French translation of *The Origin of Species* for over a decade, excitedly affirmed her devotion to a new, rational religion of progress introduced by natural selection. Royer's writings were not particularly influential, but they exemplify early social Darwinism and the way in which it utilized evolution to justify preconceived

---

68 See Robert C. Bannister, "'The Survival of the Fittest is Our Doctrine': History or Histrionics?," in *Herbert Spencer: Critical Assessments*, ed. John Offer, vol. 2 (New York: Routledge, 2000), 181–82.

69 Linda L. Clark, "Social Darwinism in France," *The Journal of Modern History* 53, no. 1 (1981): D1032.

70 See Herbert Spencer, *The Man versus the State: With Six Essays on Government, Society, and Freedom* (Indianapolis, IN: Liberty Classics, 1981), 31–34.

political, economic, and racial perceptions. Like Spencer, and influenced by Thomas Malthus, Royer sees all social relationships as governed by the economic logic of quid pro quo, where one person aids another only for the sake of reciprocal gain. In such an economic logic, value is determined by the moving equilibrium of the market. After all, the success of one individual in making money is generally premised upon the failure of others to do the same. The free market's natural flow of capital thus guarantees a just order, and the function of government is nothing more than to safeguard the market's freedom.

Social Darwinism thus originated as a new rhetoric for the old laissez-faire capitalism, newly backed by the appearance of scientific rigor. Spencer and Royer lambaste the institution of social welfare as contrary to the natural order. The law of struggle must do away with "that imprudent charity and blindness by which our Christian era has always sought the ideal of social virtue. . . . It ends up sacrificing that which is strong to that which is weak, the good to the bad, beings who are well in body and in spirit to beings who are depraved and sickly."[71] Welfare programs encourage laziness and injure the wealthy through unjust taxation.[72] The promotion of socioeconomic equality is inimical to the natural order, which uses inequality to weed out inferior human beings.

Such social Darwinism thus affirms the divisive factors of wealth, privilege, and power as natural indicators of the superiority of certain persons—white, European persons in particular.[73] The maintenance of economic and social prosperity achieves the status of a moral precept: "We will now have an absolute criterion for judging what is good and what is bad from the moral point of view," says Royer, "for the moral rule for all species is that which tends toward its own preservation."[74] Adolf Hitler will agree: "By means of the struggle the elites are continually renewed. The law of selection justifies this incessant struggle by allowing the survival of the fittest. Christianity is a rebellion against natural law, a protest against nature."[75]

Of course, most social Darwinists would not approve of genocide as a means of "purifying" the human race. However, as an ideology, social

---

71 Clémence Royer, "Préface de la Première Édition," in *De l'origine des espèces par sélection naturelle, ou Des lois de transformation des êtres organisés*, by Charles Darwin, 3rd ed. (Paris: V. Masson et fils, 1870), lxv.
72 Spencer, *Man vs. State*, 31–34.
73 See Gould, *Ever Since Darwin*, 36–37.
74 Royer, "Preface de la Premiere Edition," lxx.
75 Midgley, *Evolution as a Religion*, 139.

Darwinism has often been interwoven with dreams of population control. Darwin's dependence on Malthus's population principle made this transition natural.[76] Following Malthus, Royer argues that superior humans should be encouraged to breed and inferior ones discouraged. Darwin himself hints in this direction in his later work.[77]

Nevertheless, a critical transition occurs whenever these largely abstract aspirations for the human population are codified into concrete policies for population management. Darwin's cousin Francis Galton provided such a transition in 1869, arguing for state policies of selective breeding on the basis of hereditary traits. He later coined the word *eugenics* to describe this "science of improving stock . . . to give to the more suitable races or strains of blood a better chance of prevailing speedily over the less suitable."[78] Galton leaned in the direction of *biological determinism*, insisting that heredity played the majority role in key human qualifiers such as work ethic or intelligence. While he did not go so far as to advocate genocide, he saw welfare as the practice of preserving "sickly breeds for the sole purpose of tending them," like breeding foxes to be hunted.[79] As Galton's eugenics shows, social Darwinism was alternately utilized to support both individualism and collectivism, both the survival of the fittest individual and of the fittest genetic group.[80]

It is easy nowadays to dismiss the varied forms of social Darwinism as false and dishonest misappropriations of scientific evidence. Despite having been used primarily to back attitudes of social reform, social Darwinism accomplishes only the self-interested justification of the racist and exploitative policies of the status quo.[81] Nevertheless, this critique of social Darwinism is also too convenient. If social Darwinism's assumptions about nature and the good are grounded in culturally constructed biases, are our own ideas about good and evil entirely immune from this critique?[82]

---

76 Weikart, "Origins of Social Darwinism," 474; James Allen Rogers, "Darwinism and Social Darwinism," in *Herbert Spencer: Critical Assessments*, ed. John Offer, vol. 2 (New York: Routledge, 2000), 151.

77 Darwin, *Indelible Stamp*, 1053–54.

78 Francis Galton, *Inquiries into Human Faculty and Its Development*, 2nd ed. (London: J.M. Dent, 1911), 17 n. 1; Bowler, *Evolution*, 256–58, 308–10.

79 Galton, *Inquiries into Human Faculty and Its Development*, 18–19.

80 See Weikart, "Origins of Social Darwinism," 469–71.

81 Bowler, *Evolution*, 2; Weikart, "Origins of Social Darwinism," 473–74.

82 See Raymond Williams, "Social Darwinism," in *Herbert Spencer: Critical Assessments*, ed. John Offer, vol. 2 (New York: Routledge, 2000), 187.

In fact, Friedrich Nietzsche takes his evolutionism to such an extreme that it is questionable whether the critique of bias even applies in his case. Bias implies that one has, without proper self-knowledge, allowed preconceived notions to function as unquestioned axioms. For Nietzsche, however, the survival of the fittest within human society does not require any *a priori* or universal truth. Good and bad are themselves generated by this competition. "What is good?" Nietzsche writes, "—Everything that enhances people's feeling of power, will to power, power itself. What is bad?—Everything stemming from weakness."[83] Thus, Nietzsche can hardly be accused of being dishonest. The elites are not in power because of some pretense about innate and *a priori* superiority, but solely because of their relative success in the struggle for existence.

In the end, it does not matter that such generalized, prescriptive views on evolution may have been far from what Darwin himself intended.[84] Exculpating Darwin does not get around the real theological issues proposed by the reception of evolutionism into our cultural consciousness. In the first place, given the broader philosophical context in which Darwin's idea took root, it was inevitable that evolution would become a metanarrative of the highest order. This destabilized the metanarrative of Christianity and its concomitant anthropocentrism, but it provided nothing better than the principle of competition to fill the void of meaning left in its wake.

The advent of Darwin's theory led to a profound conflict over the meaning of the world and of humanity as its central representative. Nietzsche recognized Christianity as presenting the most dangerous challenge to social Darwinism's competitive economy of meaning. Only Christianity's sacralization of weakness, humility, and self-sacrifice can destabilize Nietzsche's nihilistic interplay of good and evil.

CONCLUSION

When we deal with evolution from the point of view of theology, we are considering something much bigger than the scientific merits of what Darwin

---

83 Friedrich Wilhelm Nietzsche, *The Anti-Christ, Ecce Homo, Twilight of the Idols, and Other Writings*, ed. Aaron Ridley and Judith Norman, trans. Judith Norman (Cambridge: Cambridge University Press, 2005), 4 (§2).

84 On the complexity of this issue, see Rogers, "Darwinism and Social Darwinism," 151.

taught. In a very real way, despite his best intentions, Darwin not only provided a new basis for biological science but also a new structure of symbolic authority. That is, since Darwin's time it has become commonplace to appeal to the theory of evolution as authorizing a variety of views that really have little or nothing to do with natural selection. The appeal to science provides an air of empirically-proven certainty. And yet, in many cases these ideas are really grounded on philosophical and religious commitments from before the advent of evolutionary science. In this sense, evolution has never functioned on the societal level as nothing but a scientific theory. Rather, evolution as it is often discussed is a grand idea, a metanarrative that purports to explain reality as a whole.

In confronting the meaning of evolution today, the key challenge facing Christian theology is how to respond to this metanarrative. As Joseph Ratzinger (now known as Benedict XVI) points out, "when natural science becomes a philosophy, it is up to philosophy to grapple with it."[85] In fact, evolutionism is more than a philosophy. For many thinkers, it is already in itself a theology. Especially for those who are influenced by process theology, the metanarrative of evolution speaks not only about the nature of the world in relation to God but even about the very nature of God's own self. Contemporary theology cannot evade the need to listen to, examine, appropriate, and critique the claims of theological evolutionism. Moreover, whether we accept or reject these claims, whether we subscribe to visions of progress or maintain Darwin's dysteleological doctrine, we must nonetheless face the question of how this response squares with Christianity and its own insistence upon explaining reality as a whole.

As we will see next chapter, the most influential evolutionary theologians of the twentieth century dealt with this problem by incorporating the metanarrative of evolution into a basically Christian framework of progress and heavenward ascent. Figures like Pierre Teilhard de Chardin did not merely see science as a confirmation of a Christian bias but allowed this science to reshape and illuminate traditional Christian notions in a way that shed light on contemporary problems and experiences.

Nevertheless, beginning in the latter half of the twentieth century, evolutionary theology will take on dramatically different forms. Returning to

---

85 Stephan Otto Horn and Siegfried Wiedenhofer, eds., *Creation and Evolution: A Conference with Pope Benedict XVI in Castel Gandolfo*, trans. Michael J. Miller (San Francisco: Ignatius, 2008), 10.

a more Darwinian standpoint, contemporary evolutionary theologians will largely undercut the anthropocentric bias of earlier pioneers and opt for even more expansive views shaped by concerns for ecology and theodicy. At the same time, the Hegelian paradigm, transmitted through process theology, will empower these theologians to speak more directly about the nature of God as an integral part of the very process of evolution. Moreover, developments such as sociobiology and evolutionary psychology will lead theologians to discuss new ways of conceiving the origin of human evil and the meaning of the human community.

CHAPTER TWO

The Birth of Evolutionary Theology
Pierre Teilhard de Chardin

The Darwinian revolution seemed at first to spell an end to the anthropocentric vision of the world that had dominated the Western mind. Humanity had esteemed itself the crowning jewel of creation because of its apparently exceptional nature. Now, it seemed that this nature was nothing more than the chance offshoot of an extinct subset of proto-apes, part of the same, broad story of biological development that linked together all of earth's organisms. But a good Idea dies hard. In 1914, even as Nikolai Berdyaev lauded Darwin's death blow to "naturalistic anthropocentrism," he retreated to a new and even more starkly defined anthropocentric paradigm: *"Man's infinite spirit claims an absolute supernatural anthropocentrism: he knows himself to be the absolute center—not of a given, closed planetary system, but of the whole of being, of all planes of being, of all worlds."*[1] In short, humanity has received a promotion: we are now the center of the universe more than ever before.

Berdyaev's response illustrates the complexity of early theological appropriations of evolution. Even as Darwin put the old anthropocentrism to death, his broader audience saw something embedded in the very concept of science that could not help but resurrect it again in a new, spiritual body. After all, it was humanity's brilliance that was finally able to decode the mystery of its own formation. "Man is the being who tells himself who he is," writes Daniel

---

[1] Nikolai Berdyaev, *The Meaning of the Creative Act*, trans. Donald A. Lowrie (New York: Collier, 1962), 73.

Day Williams.² In this sense, evolution actually became the confirmation of the kind of cosmic anthropocentric perspective found in figures such as Georg Wilhelm Friedrich Hegel; humanity was the place in which nature—characterized by evolution—at last became conscious of itself. "And it is he, by his emergence and existence, who finally proves the reality and defines the trajectory—'the dot on the i,'" writes Pierre Teilhard de Chardin (pronounced PYARE Tey·YAR de Shar·DAN).³ Humanity is nothing less than the spiritual coming-of-age of the world, "the axis and leading shoot of evolution."⁴ The transcendence of humanity is thus no longer founded on being an exception to the overarching evolutionary paradigm but rather on being the sum total and exemplary product of that paradigm itself. This is not to say that the world exists solely for the sake of human use and consumption. Nevertheless, humanity plays a unique and irreplaceable role in the fulfillment of the evolution of the cosmos.

Written between 1919 and 1955—though often published much later—Teilhard's works are emblematic of this shift to a new anthropocentrism and thus also of the birth of the field of contemporary evolutionary theology as such. He was not the first to consider the benefits of integrating the nascent science of evolution into a theological framework. He was, however, evolutionary theology's most important and influential early figure, and his ideas continue to shape the field today. Teilhard took cardinal elements visible within earlier, pioneering forms of evolutionary theology and formed these into a compelling and cohesive framework.

When examined in light of this context, Teilhard's theology thus illustrates many central features of evolutionary theology. The successors of Teilhard will variously accept, adapt, anesthetize, or abrogate these features. Some will reject his more provocative claims, which tested (if not transgressed) the

---

2 Daniel Day Williams, "The Prophetic Dimension," in *The Uniqueness of Man*, ed. John D. Roslansky (Amsterdam: North-Holland, 1969), 139.

3 Pierre Teilhard de Chardin, *The Future of Man*, trans. Norman Denny (New York: Doubleday, 2004), 58; see also *Building the Earth*, trans. Noël Lindsay (New York: Avon Books, 1969), 50, 75.

4 Pierre Teilhard de Chardin, *The Phenomenon of Man*, trans. Bernard Wall (New York: Harper and Brothers, 1959), 36. The declining status of anthropocentrism today has led some readers to argue that Teilhard is not anthropocentric at all, e.g., Karl Schmitz-Moormann, "The Future of Teilhardian Theology," *Zygon* 30, no. 1 (March 1, 1995): 126. Such an argument is possible however only by redefining the term "anthropocentrism" to cover only the most strict and archaic forms.

limits of Catholic orthodoxy. Others will not only adopt these claims, but take them ever further, holding Teilhard's challenge to Catholic dogmatism as one of Teilhard's most valuable contributions. In these ways and others, a vast portion of contemporary evolutionary theology traces its themes, language, concepts—those that are upheld, and those that are rejected—to the influence of Teilhard de Chardin.

In order to provide as deep and dynamic a view as possible within a brief space, we will examine Teilhard's ideas in view of this nexus of relationships. First, a brief glance at early forms of evolutionary theology that followed on the heels of Darwin's *The Origin of Species* will highlight the fluid intellectual heritage that fostered Teilhard's thought. Such early voices were far from monolithic, but Teilhard found within them fruitful tendencies toward organicism and vitalism, which he cultivated into a cohesive, all-unifying perspective. Second, examining many of Teilhard's key points and concepts, we will highlight important themes that will continue to shape evolutionary theology in the future. While the full extent of his theology cannot be explored here, a few important aspects will illuminate both the ways in which Teilhard was an expression of Catholic tradition and the ways in which he tested the limits of that tradition or even directly sought to contravene its then-dominant voices. Last, we will bring together many of Teilhard's key ideas in order to show how they continue to influence later thinkers. In particular, Teilhard's evolutionary theology is decisively Christocentric—focused on Jesus Christ as the ultimate fulfillment of humanity and the world. This emphasis will become an identifiable dividing line among evolutionary theologians.

CHRISTIAN ADAPTATIONS TO NATURAL SELECTION AT THE TURN OF THE CENTURY

From the publication of *The Origin of Species* in 1859 and into the early twentieth century, evolutionary theology was already taking shape. The rise of evolutionary science, combined with the popular influence of evolution in general, led to important and often inevitable theological adaptations to the challenges posed by the shifting paradigms of human meaning. In fact, while some Christians reacted with suspicion against the perceived materialism and atheism of Darwin's view, many Christian intellectuals refused to label natural selection as inherently hostile to faith. On the contrary, they found it relatively easy to bridge the gap between Darwin's dysteleological natural selection and a Christian purposive, teleological idea of development under divine guidance. This jump was not made by one monumental thinker in particular. Rather, it became a relatively common occurrence as old ideas about development adapted themselves to reinterpret—or even co-opt—the process of natural selection.

Evolutionary theology first began here in this encounter between Christian convictions about a purposeful world and a growing cultural-philosophical ethos of progress. In other words, the latter half of the nineteenth century witnessed a rush to put forth a definitive interpretation of the human *meaning* of biological evolution. Where some saw natural selection as the empirical confirmation of social Darwinism and others hailed it as a call to arms for materialism or even atheism, many Christians strove to characterize this conceptual revolution as a deeper revelation of the Creator God and God's intimate relationship with the world. In doing so, they adapted and reinvigorated existing philosophical and theological concepts, not least of which were natural theology, organicism, and vitalism.

Natural Theology: From Product to Process

The growing popularity and cultural significance of English Protestant *natural theology* during Darwin's time made such adaptation especially important. This school of theology, which reached a height in William Paley's 1802 *Natural Theology: Or, Evidences of the Existence and Attributes of the Deity*, sought to uncover observable aspects of the world that of their very nature pointed beyond themselves and toward a transcendent Creator.[5] Paley by no means represents the whole of natural theology, yet his work held a lasting influence on Darwin and his English-speaking contemporaries.[6] The depth of Paley's scientific observations, coupled with the force of his arguments in favor of design, commanded Darwin's respect and admiration despite his eventual rejection of Paley's point of view. For Paley, the intricate complexity of creatures both in terms of their structures (e.g., the eye) and in terms of their relations to their environments (e.g., the perfect correspondence between the length of the hummingbird's tongue and the flowers from which it drinks)

---

5 McGrath, *Darwinism*, 12, 161.
6 Bowler, *Evolution*, 38–40; Darwin, *Autobiography*, 59.

could only have been designed by an intelligent and providential Creator. In view of this, empirical science actually serves as apologetic function. By highlighting the beautiful complexity of creatures, it testifies to the meaningful and providential relationship between God and the world.

Far from eliminating natural theology, Darwin's discovery gave new impetus to its ideas. Major defenders of Darwin, such as the American botanist Asa Gray (1810–88), reinterpreted natural selection as providing evidence of purpose and design in creation.[7] In a sense, Gray's arguments represented a retreat position. He had to reconceive design less in terms of direct, concrete divine intervention in the development of an individual species and more in terms of a broader principle of design evident in the very process of natural selection. Nevertheless, Gray held fast to the optimism of Paley's position, arguing that the omniscience of divine wisdom would have foreseen and thus intended the outcomes of natural selection. Darwin's idea thus evidenced the genius of an act of creation that could establish a desired end by generating perfect biological laws at the beginning.

The primary accomplishment of such theological adaptation was a shift in focus from product to process. Paley's view had located evidence of design, especially in the finished product of creation. For the Anglican priest Charles Kingsley (1819–75), this represented an overly static view. Thus, for Kingsley, Darwin had actually rescued natural theology by putting an end to this limitation.[8] Now it was theology's turn to rescue evolution from the specter of materialism. Kingsley accomplished his goal by combining the meaning and purposiveness of creation with the dynamism and progressive character of evolution. Creation represented an ongoing and dynamic process in which divine providence imbues creatures with the capacity to realize new possibilities. The result of evolution is not entirely predetermined beforehand, and yet it is still meaningful as the product of the ongoing process of creative divine agency embedded within nature itself.

Such a renewed theological emphasis represented both a revitalized view of creation and a dynamic understanding of the Creator. In opposition to the God of Deism, who creates the world and then leaves it to its own devices, post-Darwinian natural theology continued to develop an image of God as intimately involved in the world in a dynamic, ongoing way. Even if particular adaptations occur through the mechanism of natural selection, this

---

7 Livingstone, *Darwin's Forgotten Defenders*, 63–64; McGrath, *Darwinism*, 162–63.
8 McGrath, *Darwinism*, 164.

does not reduce God's involvement but actually deepens it. God's continued providential governance shines forth within the progressive development of species. Thus in 1903, the theologian James Orr argued that evolution could be reconciled with Christianity if God's intervention is not understood as external meddling but rather as something internal to the world and its processes. "The real impelling force of evolution," he writes, "is now from *within*; it is not blind but purposeful. . . . Evolution is but the other side of a previous *in*volution and only establishes a higher teleology."[9]

Such accommodative natural theology represents only one of a plethora of early Christian responses to evolution. Nevertheless, it established many of the key strategies that will variously shape the evolutionary theologies of the twentieth and twenty-first centuries. For later theologians, if evolution is truly to get along with Christian theology, it is not enough for it to be merely compatible with doctrine. Rather, evolution must actually reveal something deeply true, providing a unique perspective that highlights the intimate truth within theology itself.

The Revival of Organicism

By challenging an unimaginative, mechanistic view of creation, evolution also reinvigorated a Christian view of the world as a living, dynamic organism. Thus in 1889, the Anglo-Catholic theologian J. R. Illingworth praised evolution's attack on anthropocentrism as one of its key theological contributions. Humanity's complaisant centrality had led past thinkers to see the unity and design of nature principally in terms of its usefulness to human beings: "They saw the utility before the beauty . . . as if corn should exist solely to provide food for man."[10] According to Illingworth, such a utilitarian view resulted in an overly mechanistic view of nature. Treating nature as a colossal machine, a mechanistic view reduces the meaning of the natural world to its function. In relation to this, the various creatures of the world are only meaningful to the extent to which they support the machine's functioning. If, for example, it is still able to function without a few of its gears, then each gear is in itself meaningless, replaceable, and thus disposable. After all, a gear is a means and not an end in itself.

---

9 Orr, *God's Image*, 96.
10 Illingworth, "Incarnation," 138.

For Illingworth, evolution's critique of anthropocentrism thus helped theology to recover a more organic concept of the world. If the purpose of the world is not to serve the whims of humanity, and if humanity is moreover a part of that world, then the world's purpose must be to develop its own inner perfection for the good of its varied parts. In this way, the world starts to look less like a machine, which serves some external purpose, and more like a living organism, which is meaningful in its own right, without reference to any external goal.

Illingworth thus illustrates the early rise within evolutionary theology of the tendency to perceive the world in terms of *organicism*. Rooted in the unitive worldview of Platonism as well as Paul's frequent biblical depictions of the Church as the body of Christ (e.g., 1 Cor 12:12-31), organicism interprets the world and its creatures by analogy to the body and its members. The world is one living, organic body composed of different but interrelated parts, all of which contribute to the good of the whole. At the same time, the whole always remains more than the sum of its parts. Just as a human is more than a head with a torso, arms, and legs attached, so also the world. Each part of the body has its own proper meaning and cannot simply be replaced by another. Thus the brain is vital for the organism, but a brain is no suitable replacement for a leg. The brain needs the legs even as the legs need the brain. Hence, each part realizes its own proper meaning most perfectly in connection with the organic whole. As the unity of the body enables the brain to fulfill its purpose, so also does the unity of the world highlight the real meaning of the various creatures within it. In this way, each individual being is both a means *and* an end. Each organic part is both an end *in itself* and a means to a higher, transcendent end, an end which is expressible only on the level of the entire organism taken together.

In contrast to the utilitarian structure of meaning implied by mechanistic thinking, organicism typically appeals also to a sense of beauty and wholeness. While corn is indeed meaningful for the way in which it provides nourishment to humans and others, it is also meaningful and *beautiful* in its own right. Plato's broad influence on Christianity stands behind this assumption. For Plato—and even more so for the Neoplatonist Plotinus—the variety of creatures within the world necessarily represents the transcendent beauty of the divine Ideas. The world must represent this variety in its entirety or else it would fall short of its own meaning or purpose. For Illingworth, accordingly, the discovery of evolution actually reveals the beauty of this world. The very facts of nature uncovered by scientific inquiry serve to illuminate the profound harmony that marks the graceful movements of an organic body: "We discover the quality of beauty in every moment and situation of this complex life; the drop of water that circulates from sea to cloud, and cloud to earth, and earth to plant, and plant to life-blood, shining the while with strange spiritual significance in the sunset and the rainbow and the dewdrop and the tear."[11]

This sense of beauty and wonder imbues such theology with the feeling that, even if nature is expressible as Darwin's "struggle for existence," even if it is forever locked in an evolutionary process of becoming something other, it is always already at the same time something more, something meaningful. In the words of Illingworth, each creature is "at once a revelation and a prophecy, a thing of beauty and finished workmanship, worthy to exist for its own sake, and yet a step to higher purposes, an instrument for grander work."[12]

Metaphysics: Vitalism and Creativity

This quest to integrate Darwinian evolution into a purposeful view of the world was not restricted to the realm of Christian theology. As we have seen, the tradition of metaphysics provided ample impetus to reintroduce a sense of progress into natural selection. From Plato to Christianity, refracted through Leibniz and others, a mass of underlying philosophical influences had driven thinkers like Spencer to conceive of the world as a unified system propelled in a particular, positive direction by some inner spiritual force.

Such a metaphysics of evolution took hold alongside Darwinian natural selection in part because of a reluctance among many readers—for scientific more than for religious reasons—to accept Darwin's insight in isolation. Instead, the rise of Darwin's theory of natural selection helped to spur greater appreciation for the ideas of Jean-Baptiste Lamarck (1744–1829).

In contrast with natural selection, Lamarckian evolution focused heavily on the active role of the individual organism. Lamarck was convinced that environmental changes led to biological modifications. Darwin borrowed and expanded this claim. However, whereas Darwin wanted to see these external, environmental factors as sufficient of themselves, Lamarck was

---

11 Illingworth, 139.
12 Illingworth, 139.

reluctant to attribute the brilliant adaptation of species to objective causes alone. Evolution, he believed, is not something the environment does to the organism so much as something the organism does in response to the environment.[13] Individual organisms adapt to environmental changes by developing new habits and emphasizing the use of certain organs or faculties. Such shifts in habit and activity produce physiological changes, just as frequent jogging strengthens the legs. If both the male and the female of a mating pair happen to develop the same modification, Lamarck insists, their offspring will be born already modified. To illustrate: If a man and woman are stranded on a desert island and become expert swimmers as a result, their child will be born with this expertise already in place.

Given the ease with which contemporary science refutes Lamarck's most distinctive claims, people today often regard his theories as archaic, obsolete, or even comically absurd. Yet this convenient and cavalier attitude toward Lamarck was only made possible in the mid-twentieth century by the resurgence of a more authentic Darwinism via the progressive integration of Gregor Mendel's laws of inheritance and the development of the science of DNA.[14] Prior to what Julian Huxley in 1942 termed "the modern synthesis,"[15] Darwin's inability to identify how mutations arise in the first place or to fully account for the establishment of non-beneficial traits led many scientists and philosophers to accept evolution in general while looking elsewhere to fill in natural selection's apparent gaps. Whereas Lamarck's ideas had been controversial early on in the nineteenth century, the evolutionary fervor incited by Darwin's *The Origin of Species* sparked new interest and appreciation. In the United States, for example, Edward Drinker Cope and Alphaeus Hyatt developed an approach that would become known as neo-Lamarckism. Their attempts to fit natural selection within an overarching framework of intelligence and higher purpose came to rely upon an

---

13 Jean Baptiste Lamarck, *Zoological Philosophy: An Exposition with Regard to the Natural History of Animals* (New York: Hafner, 1963), 112.

14 Gregor Mendel published his groundbreaking findings on inheritance in 1866. In 1882, Walther Fleming first described the division of chromosomes within cells and coined the term "mitosis." It was not until 1902, however, that Walther Sutton and Theodor Heinrich Boveri, working independently, recognized DNA as the physical mechanism behind Mendel's laws of inheritance. Jean Vallade, "The Slow Emergence of the Meiosis Concept from 1882 to 1909," *Acta Botanica Gallica* 160, no. 1 (2013): 4, 7–8.

15 Julian Huxley, *Evolution: The Modern Synthesis* (New York: Harper and Brothers, 1942).

essentially Lamarckian basis.[16] As one of Darwin's American supporters quipped, "It would seem, at first sight, that Mr. Darwin has won a victory, not for himself, but for Lamarck."[17]

Darwin's own openness to Lamarckism grew over time. This no doubt contributed to its symbiotic growth. In his first edition of *The Origin of Species*, while Darwin saw natural selection as the central mechanism of biological evolution, he accepted Lamarck's theory of inheritance by way of habit as a lesser, secondary contributor. Yet as his theory matured, he came to allow Lamarckian inheritance—along with sexual selection and unexplained mutations—a more significant role.[18] This shift permitted moderates and Darwin's more apprehensive supporters to accept natural selection not as the exclusive or even primary means of biological evolution but rather as one key mechanism among others.

Yet Lamarckism also offered something more than a mechanism of change. For many, it provided an integrative framework for binding natural selection and other developmental mechanisms into an overarching system of meaning. By privileging the active role of the organism, it enabled one to see evolution as a process of self-improvement or self-transcendence. It reopened the possibility of seeing this process as the unfolding of a higher purpose within the world. The adaptive action of the individual organism could be seen as proceeding from an inner and invisible drive or force that propels evolution toward its definitive goal. In short, Lamarckism allowed theorists to wrap the empirical facts of evolution within a synthetic, metaphysical system that acted as a layer of cosmic meaning.

In this way, especially for those coming from a metaphysical or religious standpoint, Lamarckism helped to dull the sharp edge of natural selection. It brought back a sense of purpose, design, and progress, which did not fit within a strict Darwinian paradigm. For Lamarck, the orderly adaptation of organisms to the environment stood as evidence that evolution must be governed from within by an innate vital principle or life force.[19] Although he accepted the materialist view of the body as a kind of machine, he believed

---

16 Bowler, *Evolution*, 242; see also "Edward Drinker Cope and the Changing Structure of Evolutionary Theory," *Isis* 68, no. 2 (June 1977): 249–65.

17 Wright, *Darwinism*, 5.

18 Darwin, *Indelible Stamp*, 416–22; *Autobiography*, 49, 152–53; see Yongsheng Liu, "Darwinian Evolution Includes Lamarckian Inheritance of Acquired Characters," *International Journal of Epidemiology* 45, no. 6 (December 1, 2016): 2206–2207.

19 Livingstone, *Darwin's Forgotten Defenders*, 54.

that the movement of this machine must be initiated by a vital principle that, while not supernatural, nonetheless transcends the merely mechanical and receptive character of the body.[20] In view of this, both the individual and the species could be seen as having some transcendent, purposeful meaning.

In short, the scientific merits of Lamarckism permitted a metaphysical *vitalism* to survive the implicit critique leveled against such thinking by Darwin's natural selection. Though it takes many diverse and competing forms, vitalism is the idea that a transcendent, universal, nonphysical principle or life force drives the history, development, and activity of organic life.

Despite its religious and sometimes even mystical overtones, vitalism developed out of materialist attempts to extricate the action of God from the apparently intelligent, creative, or artful operations of the natural world. The rise of Newtonian physics in the eighteenth century led thinkers such as Pierre Louis Moreau de Maupertuis (1698–1759) to insist that even the biological world must function on mechanistic principles. Maupertuis thus rejected the then-common assumption that an embryo forms in the womb according to a preexisting and divinely established pattern. Yet as this was long prior to the science of DNA, this denial left him without any clear means of explaining the marvelous complexity produced by the embryo's formative process. Maupertuis's attempt to fill this gap relied upon the supposition of an invisible, omnipresent, and objective force—something akin to Newton's gravity. An apple falls to the ground not because it obeys a transcendent divine pattern but rather because its material—its mass—is intrinsically governed by the invisible draw between the mass of the earth and the mass of the fruit. Maupertuis averred that organic matter must have its own inner force, which from within governs its development toward a complex, well-ordered, and predictable pattern—the species. In this way, Peter J. Bowler notes, "in attempting to explain life as the product of material processes, philosophers were forced to attribute the properties of life to matter itself."[21]

This materialist appeal to vitalism did little to clarify the complex and well-ordered character of organisms. It removed God from the equation, but as a suitable substitute it provided nothing more than a vague, inner orientation toward self-arrangement. It avoided the all-too-Platonic assumption that species accord with universal patterns preestablished in the mind of God, but it simply injected this patterning function into biological matter itself.

---

20 Lamarck, *Zoological Philosophy*, 236.
21 Bowler, *Evolution*, 74–75.

Despite these problems, subsequent thinkers continued to feel the impact of materialism's critiques against relying on divine ordination for the operation of natural processes. Figures such as Lamarck adopted new forms of vitalism as a means of understanding how the development of organisms toward order, complexity, and intelligence can belong to the organism itself without any strict, external divine control. Despite its origins, Christians did not necessary see such a vitalist approach as inherently atheistic. God could still be understood as the author of nature and of the vital principle itself without having to directly govern and maintain the orderly arrangement of organisms and species. A pure Darwinian natural selection would eliminate the need for such a vital principle by making adaptation inherently *unintelligent*—mechanical—but prior to "the modern synthesis," even many of Darwin's supporters held on to forms of vitalism into the twentieth century.

Henri Bergson's pivotal 1907 work *Creative Evolution* highlights the critical role that revamped forms of vitalism came to play in the development of evolutionary theology. A Jewish philosopher with professed Catholic sympathies, Bergson sought to conceptualize the world as evolving from within without reducing this development to a mechanical or mechanistic process. Creativity was key to his approach. A machine creates nothing, because it cannot do anything *new*. It merely performs the operation for which it was designed, so that its range of possibilities is strictly determined beforehand. A wooden abacus cannot connect to Wi-Fi without a fundamental change in its basic design.

Many metaphysical and materialist views alike restricted the creativity of evolution by constraining its movement to predefined parameters. On one extreme, the traditional metaphysical principle that "like generates like" fed into views that saw evolution as nothing more than a gradual rolling-out of possibilities according to predefined divine patterns. Creatures develop in a specific and predictable way, because what the creature is to become is already latent in what it is.[22] Just as a tulip bulb will only become a tulip, so also was the diminutive eohippus fated to become the modern horse. On the other extreme, T. H. Huxley's materialist approach saw evolution as utterly predetermined by the prior, accidental situation of material elements. A bowling ball might spin a certain way because of the tiniest variation in the smooth lane or the slightest imbalance in its formation. In the same way, Huxley saw

---

22 Henri Bergson, *Creative Evolution*, trans. Arthur Mitchell (New York: The Modern Library, 1944), 34.

every concrete fact as essentially predetermined by past conditions even so distant as the random arrangement of the "cosmic vapour" that, to his mind, preceded the galaxies; as a consequence:

> A sufficient intelligence could, from a knowledge of the properties of the molecules of that vapour, have predicted, say the state of the Fauna of Britain in 1869, with as much certainty as one can say what will happen to the vapour of the breath in a cold winter's day.[23]

Huxley saw the essential predictability of all material processes as a benefit of this materialist approach.

For his critics, Huxley's view epitomized the fatal lack of genuine creativity within pure Darwinian evolution. It made present-day species nothing more than the mechanical result of earlier, primordial happenstance. Could blind, impersonal, and unintelligent mechanisms be wholly responsible for the dramatic changes whereby a microbe became the eons-distant ancestor of an elephant? To the Scottish American theologian Ebenezer Nisbet, such materialistic Darwinism seemed like an endless struggle to explain the appearance of new species by seeking out yet another ingredient to add to the mysterious recipe. For Nisbet, a devoted Darwinist was like a madman stirring a saucepan of assorted pig parts, believing that just one more ingredient will bring the pig back to life. We have "everything that is necessary," declares the madman, "we only lack the vital warmth, and the young pig will be ready made again." Nisbet writes, "Darwinism is forever like the lunatic, looking for his pig to appear."[24] Without some divine creative principle, how could life accomplish any genuine movement into something new?

Without creativity, without the arrival of something genuinely new, what really separates a species from the conditions that spawned it? What makes a human anything more than a glorified ape? Gilbert and Sullivan mocked this idea in their 1884 opera *Princess Ida: Or, Castle Adamant*. Ida's dislike of men is explained using a parable about an ape who falls in love with a beautiful woman. A symbol of "Darwinian man," the ape shaves himself and dresses in fine clothes to impress his love. Nevertheless, he is doomed to fail in the end,

---

23 T.H. Huxley, "The Genealogy of Animals," in *Critiques and Addresses* (London: Macmillan, 1873), 305–306.
24 Ebenezer Nisbet, *The Science of the Day and Genesis* (C. Venton Patterson, 1886), 67–68.

for "a Man, however well-behaved / At best is only a monkey shaved!"[25] Are humans are nothing more than the sum of their ancestors? Humanity must be the same old ape with a new haircut, since genuine creativity requires the arrival of something authentically new.

For historical Christianity, the concept of divine creation had provided the requisite sense of newness. According to patristic interpretations of the book of Genesis, God created the world out of nothing (*creatio ex nihilo*). Going from the boredom of nothing to the novelty of something, this creation of the world is the arrival of absolute newness. Such newness is premised upon the divine imagination, which alone is able to form an idea of something truly new and unanticipated. In view of divine creativity, the future can be both foreordained by God and yet authentically malleable and open from the human point of view. Nevertheless, for many modern thinkers, this Christian emphasis on creation makes creativity too extrinsic to creation itself. The genuine newness of the world derives in an active sense from the creativity of God. The world's potential for creativity—its foundation upon nothingness—represents a passive, receptive potentiality. It is a blank canvas waiting for the artist's skilled hand.

In contrast, the heterodox vitalism of John Fiske (1842–1901) unites divine creativity and the world's inner creativity in an absolute manner. A supporter of Darwinian evolution and a disciple of Herbert Spencer, Fiske insists that the vital principle behind all worldly development is itself divine; it is the very creativity of God. In short, Fiske's vitalism is a form of *monism*, a theory that reduces all reality—even the reality of God—to the progressive action of a single fundamental principle. While such a view seems to provide the world with a dramatic sense of freedom and creativity, it actually reduces this to a kind of divine determinism. After all, a monistic principle is like a single mathematical algorithm that produces both God and the world in a necessary and mechanical manner.

Bergson's philosophy is set against the background of this problem of creativity. His goal is to reintroduce a sense of openness and newness without sacrificing the mechanical processes of evolution, without relying on a merely extrinsic, divine principle of creativity, and without reducing God to a vital principle. For Bergson, the evolutionary progress of the world, even as it makes sense as the outcome of prior conditions, is yet also the

---

25 W.S. Gilbert, *Princess Ida: Or, Castle Adamant* (London: G. Bell, 1912), 27.

unpredictable product of a genuine intrinsic creativity, a newness that erupts from within the world itself. He terms this principle of creativity the vital impulse or *élan vital*.[26]

Bergson rejects the Lamarckian form of vitalism, which he sees as too individualistic. Lamarck saw the vital principle as working through the actions of heat, electrical impulses, and other such unseen factors within the individual organism. His vital principle governed the way in which the individual organism actively adapted to the environment. In contrast, Bergson's vitalism operates more on the level of the world as a whole. It is not visible in an individual raccoon's adaptation to living in an attic but rather in the overall flow of biological evolution. The vital impulse is not extrinsic or unnatural, but neither is it possessed in a direct and simple form by any single creature or species.

Bergson's *élan vital* is thus aimed not at justifying a mechanistic view of the world but rather at establishing the process of evolution as both meaningful and, at the same time, utterly free and non-predetermined. He admits with T.H. Huxley that the material, historical conditions of the world have a real impact upon evolution, but he denies that these conditions alone suffice for explaining the real meaning and direction of evolution as a whole. Bergson illustrates with the analogy of road construction. As history constructs the road (the pathway of evolution), natural selection explains its ups and downs, "but not its general directions, still less the movement itself."[27] Historical particulars, the "accidents of the ground"—its underlying hills, valleys, rocks, and rivers—necessarily shape the road, but they do not cause it to exist. Rather, the road is built only because it is *teleologically* directed—it reaches out toward an end, the town. Precisely how it arrives there does not negate this overarching and original purpose.

Bergson does believe that evolution moves ultimately in something like an upward trajectory. However, it does not proceed toward a fixed end or along a single path, and every step along the way is open to chance and accidental determination. A static, preconceived divine plan would seem to indicate a closed, predetermined future. Instead, Bergson describes a dynamic of ongoing creation, wherein the inner vital impulse realizes itself through the infinite, open possibilities of biological evolution. Evolution is going

---

26 Bergson, *Creative Evolution*, 49–50.
27 Bergson, *Creative*, 114.

somewhere, and this somewhere is foreordained by the progressive dynamism of the vital impulse itself.

Summary of Influences

The abovementioned developments will have a lasting impact on evolutionary theology and in particular on Pierre Teilhard de Chardin. Natural theology's shift from product to process will revitalize the link between God and the world by making the entire history of the world's development meaningful. This goes hand in hand with a concerted organicism, which highlights the interconnection between and meaningfulness of all creatures. In view of this, various theological views will see divine governance of the world not as an external force, but rather as something internal to the world as a living being: a kind of vitalism. Rather than being the mere product of meaningless mechanical forces, the world will be seen as evolving through the guidance of authentic creativity, which opens it up to new and exciting possibilities.

In short, Christian theology quickly colonized natural selection into a teleological form. They adapted scientific claims about evolution to fit into frameworks that viewed world history—even biological history—as progressing toward fulfillment in a higher end. Though such an approach violated or even ignored Darwin's doctrine of natural selection, it did not immediately bring Christianity into conflict with science, since many scientists of the time accepted teleology all the same. Rather, to the contrary, this adaptation forged the beginnings of a mutually constructive relationship between science and theology. Early evolutionary theologians set a pattern whereby theology would adapt to scientific critique and utilize science to deepen its theological perspective on the relationship between God and the world.

PIERRE TEILHARD DE CHARDIN: EVOLUTION CONTINUES FROM BIOSPHERE TO *NOOSPHERE*

In view of these early influences, we are now in a better position to understand the phenomenon of Pierre Teilhard de Chardin (1881–1955). A French Jesuit priest, paleontologist, and geologist, Teilhard expanded the insights of his predecessors and continued to build evolutionary theology into an all-encompassing worldview centered upon the incarnation. His particular form of evolutionary theology will exercise a dramatic influence on later thinkers and set the tone for significant developments and divergences within the field.

It would be no small feat to treat in this small space the breadth of Teilhard's corpus, the richness of his life, and the fecundity of his influence on others. Accordingly, this examination must be limited to tracing some of the key lines of thought that Teilhard draws from earlier sources, develops, and bequeaths unto subsequent thinkers. In particular, Teilhard draws a close link between organicism and the unity of all creation. Espousing a highly purpose-driven view of evolution, Teilhard views history in terms of the inner movement of matter toward spirit. The pieces of this worldview fit together beautifully. However, they form a picture of evolution that is so radiantly optimistic as to stand in stark contrast to the Darwinian idea of evolution as a "struggle for existence."

In exploring these elements, we will see how such an integrative thinker can inspire multiple distinct and even incompatible forms of evolutionary theology. This phenomenon hinges in part upon the ways in which readers consider Teilhard's more controversial statements. A statement that one reader brushes off as an incidental mistake can become for another a central feature of Teilhard's thought. Without subjecting Teilhard to a tribunal, we will briefly examine how Teilhard's forays into the problem of evil and the metaphysics of God help to situate the compatibility of some forms of evolutionary theology with process theology. At the same time, Teilhard's intense and deeply traditional emphasis on the centrality of Jesus Christ also serves as a distinguishing element in the field, so much so that the presence of Christocentrism in later forms of evolutionary theology often points to the distinctive influence of Teilhard over other sources.

The Intrinsic Direction of Evolution toward Unification

From our brief examination of post-Darwinian natural theology, one can see evolutionary theology as an ongoing development rooted in a variety of philosophical, theological, and scientific ideas. Teilhard de Chardin is no random aberration. Rather, he stands as a significant moment within this broader story of the evolution of theology. In fact, this intellectual heritage is important precisely because Teilhard himself sees all of history—including the history of human thought—as an interconnected, unified, and purposive process on the way toward higher fulfillment. Even apparently scattered and discontinuous moments—even, for example, the dispersion of human thought into various disciplines and approaches—are really part of the universe's more fundamental and all-encompassing momentum toward the realization of a deeper and more perfect sort of unity.

Nevertheless, it can be difficult to pinpoint the exact sources that fed into Teilhard's evolutionary theology. Teilhard was a man of varied experiences. He studied philosophy, theology, mathematics, literature, and science; taught physics and chemistry; carried a stretcher in World War I; participated in important paleontological digs (most notably the early hominin called Peking Man); and visited numerous countries around the world. Though clearly influenced by Bergson, Teilhard himself presents him less as an intellectual progenitor and more as a kindred spirit whose works resonated with insights already taking shape in his own mind.[28]

To some extent, Teilhard downplays the impact of earlier thinkers on his works because he understands his own intellectual development not primarily as an academic pursuit but rather as a journey of mystical intuition. From an early age, he held a profound appreciation for the beauty and harmony of the natural world. Coupled with a Catholic predilection toward universal unity and organicism, this led him to accept a deeply teleological, ideological concept of evolution. Evolution became for him "that magic word . . . which haunted my thoughts like a tune: which was to me like an unsatisfied hunger, like a promise held out to me, like a summons to be answered."[29] In short, though a scientist by profession, Teilhard's acceptance of evolution was prior to and deeper than any mere empirical, scientific confirmation. Darwin had all too little to do with it. Rather, Teilhard's was an evolutionism that, together with Catholicism, formed a coherent metanarrative ordering all of his empirical experiences toward the ultimate question of meaning. Building upon this basis, his research as a scientist served only to deepen his spiritual convictions.[30]

---

28 Pierre Teilhard de Chardin, *The Heart of Matter*, trans. René Hague (New York: Harcourt Brace Jovanovich, 1979), 25; cf. Mary Lukas and Ellen Lukas, *Teilhard* (Garden City, NY: Doubleday, 1977), 33.

29 Teilhard de Chardin, *Heart of Matter*, 25.

30 Teilhard de Chardin, *Building*, 67–68; see also *Christianity and Evolution*, trans. René Hague (New York: Harcourt Brace Jovanovich, 1971), 96. For a good analysis of Teilhard's key mystical experiences, see David Grumett, *Christ in the World of Matter:*

The central importance of Teilhard's mystical spirituality easily overshadows the philosophical and theological influences behind his ideas. Yet these influences are important as well. It is true that there is little use of Thomas Aquinas or Aristotle in his thought. At times Teilhard directly critiques this tradition, especially as embodied within neo-scholasticism, the dominant form of Catholic thought from roughly 1879 to 1950. Nevertheless, not only is Teilhard not altogether discontinuous with Thomistic thought,[31] he is in many ways deeply continuous with the even more long-standing tradition of Christian Platonism—the same philosophy that inspired figures such as Augustine, Pseudo-Dionysius, Meister Eckhart, and much of the Christian mystical tradition. This heritage underlies the significant points of contact that exist between Teilhard's thought and the Neoplatonic philosophy of Plotinus (204/5–70 CE).[32] It also grounds Teilhard's strong emphasis on organicism and meaning.

In line with nineteenth-century thinkers, Teilhard appeals to organicism to ward off the danger of a mechanistic worldview. The world is not a robot, but a living and dynamic organism. An organism resembles a machine by the interworking of its parts. Yet an organism is always also something more—something meaningful. In this view, therefore, the world takes on a new and richer meaning when viewed as a whole. At the same time, the various beings within the world are irreducible to mere cogs in a machine. They have their own intrinsic value that is further demonstrated by the unity of the whole. By analogy, the arm contributes something unique that cannot be supplanted by the operation of the leg; therefore, the arm has a meaning that shines forth in the whole, but it is not reducible to the meaning of the whole as such. The beings within the world play an integral part in the realization of the meaning of the world, but they are never a mere means to an end.

Plotinus establishes an important precedent by understanding the meaning of the world in terms of contemplation (Greek *theōriā*). In his view, the

---

Teilhard de Chardin's Religious Experience and Vision (Woodbridge, CT: American Teilhard Association, 2013).

31 See Francis J. Klauder, *Aspects of the Thought of Teilhard de Chardin* (North Quincy, MA: Christopher Pub. House, 1971), 17–28.

32 On this topic, see Mary T. Clark, "The Divine Milieu in Philosophical Perspective," *The Downside Review* 80, no. 258 (1962): 12–25; Jacobus Stefanus Oosthuizen, *Van Plotinus tot Teilhard de Chardin: 'n Studie oor die metamorfose van die Westerse werklikheidsbeeld.* (Amsterdam: Rodopi, 1974); David Grumett, *Teilhard de Chardin: Theology, Humanity and Cosmos* (Leuven: Peeters, 2005), 7–22.

innate creative power of nature (*physis*) for change or development stems from an innate connection with reason (*logos*). Physical changes are never merely dumb or irrational, for they are manifestations of the divine Ideas within matter. Accordingly, Plotinus argues, "Nature [*physis*] produces by virtue of being an act of contemplation [*theōriā*], an object of contemplation and a *Logos*; on this triple character depends its creative efficacy."[33] In other words, nature exists by way of and in order to envision the divine Ideas within itself. An expansion of Plato's conception of the unity of being and knowing,[34] Plotinus's schema thus paints the very purpose of the world as being realized in the propensity for finite beings to contemplate the beauty and perfection of the divine Ideas manifested in and through the world. This means, moreover, that intelligent creatures (i.e., humans) contribute directly to the meaning of the world by their ability to appreciate its beauty.

By mapping this kind of perspective onto an evolutionary, developmental understanding of world history, Teilhard de Chardin stretches the unity of being and knowing onto a temporal axis. The more the world *comes to be*, the more it *becomes conscious of being*. Evolution is, quite literally, "the rise of consciousness."[35] Humanity plays a central role in this schema because we are, in effect, the conscious organ of the world. We are the brain that, by evolving out of lower levels of intelligence, becomes at last conscious of itself. Far from alienating us from the rest of creation, this role further underlines the unity and dignity of finite beings as part of one world organism.

For Teilhard, the development of human consciousness effectively accelerates the process of evolution. It turns it from an apparent mechanism of differentiation and dispersal and into a process of unity through difference.[36] In this way, the oneness of the world is not merely something given beforehand, e.g., a statement of our common origin. Rather, the oneness of the world is at the same time a kind of ultimate destiny, the definitive goal and orientation of the world as such. Just as a person's individuality develops through

---

33 Plotinus, *Enneads*, trans. A. H. Armstrong (Cambridge, MA: Harvard University Press, 1966), I, 8, 3. Translation modified slightly for terminological clarity.

34 Plotinus, *Enneads* I, 8, 7–8. Plato, *Republic*, book VI. On this topic, see Grace A. De Laguna, "Being and Knowing: A Dialectical Study," *The Philosophical Review* 45, no. 5 (1936): 435–456.

35 Teilhard de Chardin, *Phenomenon of Man*, 243; *Activation of Energy*, trans. René Hague (New York: Harcourt Brace Jovanovich, 1971), 157; cf. Bergson, *Creative Evolution*, 32.

36 Teilhard de Chardin, *Future of Man*, 153–55.

adolescence and adulthood, so also does the world develop in its oneness as human consciousness expands and unfolds. Humankind is destined to oneness with each other and with the world as a whole neither by way of abstraction nor by any violent leveling of differences but rather by way of a "union of concentration (the *only* true union)," which "does not destroy but emphasizes the elements it swallows."[37] The unity of a body does not make all appendages into arms but maintains and elevates the significance of each organ for its own sake. The goodness of the arm does not negate the goodness of the leg. Just so, the evolving unity of the world intensifies the individuality and irreplaceability of the persons within it.

Teilhard frames this acceleration of unification by way of consciousness in terms of an entirely new phase or "layer" of evolution: the *noosphere*. He coined this term as a play on the word "biosphere"; just as the latter represents the biological aspect of the world, the former represents the layer of the world's conscious thought.[38] Yet this noosphere is more than the collective chatter going on within people's heads. For Teilhard, it is a concrete and distinct reality, something like "the very Soul of the Earth."[39] Just as the individual is a body who is conscious of itself, the noosphere represents the consciousness and subjectivity of the world as a collective whole. Within the noosphere, humanity continues to evolve, especially by way of the development of its more conscious aspects such as culture, morality, and spirituality. Biological evolution does not cease upon the advent of the noosphere. However, with this advent, evolution as a whole has reached a defining phase in the process of unification. Human biology may continue to develop, but only through the further intensification of consciousness can the full unification of the world be realized as voluntary and free—that is, as a unity of love. Biological development alone cannot bring about that future toward which Teilhard sets his hopes.

The noosphere is thus the final stage of a truly *cosmic* evolution—a developmental trajectory that is intrinsic to matter and which occurs on every layer of material existence. From the beginning, evolution is a principle of process that is more fundamental and expansive than biology alone. It begins at the smallest level of inanimate matter: atoms come together to form increasingly

---

37 Pierre Teilhard de Chardin, *Human Energy*, trans. J.M. Cohen (New York: Harcourt Brace Jovanovich, 1971), 104.

38 Teilhard de Chardin, *Future of Man*, 149–51. "Biosphere" was coined by Eduard Seuss in 1875 and popularized by Vladimir Vernadsky in 1926.

39 Teilhard de Chardin, *Heart of Matter*, 30–32.

complex molecules. Over a vast span of time, this precipitous complexification leads to the development of organic life. The same tendency toward complexification will eventually bring about the development of conscious life and humanity. In this way, the noosphere is not the accidental result of blind chance. It is rather the sum product of the meaningful and intrinsic orientation of matter toward spirit.[40]

The Interplay of Matter and Spirit

Just as evolution reaches from biosphere to noosphere, so also does it shape the history of both matter and spirit. Teilhard's approach thus hinges upon a metaphysics that sees matter and spirit as intrinsically related to one another. Teilhard's exploration of this relationship will have important consequences for the future of evolutionary theology.

The problem is quite simple: if the human body is a mere product of biological alteration from an earlier animal, and this same alteration gave birth to human intelligence via the complexification of the brain, then what is the status of the soul? It is easy to proceed from human evolution to materialism, the view that the world is fully understandable in view of material reality without any appeal to a transcendent spirit or soul. An orthodox Catholic approach to evolution must therefore secure some sense in which spirit is irreducible to matter, and yet if done in a way that makes an exception of humanity—e.g., evolution holds true in all cases *except* for the human mind—this would contradict the very evidence that leads us to the idea of biological evolution in the first place.

Reconciling the existence of the soul with the material reality of evolution is further complicated by the different ways in which the terms "matter" and "spirit" have functioned both in the tradition and in modern thought. An oft-repeated narrative claims that Christianity was inherently dualistic, despising matter in favor of spirit, and that Teilhard sought to overthrow such dualism and proclaim the nondual oneness of spirit and matter.[41] Yet this melodramatic account fails to do justice to Christianity's long history as the principal defender of matter and flesh against philosophical and religious

---

40 Teilhard de Chardin, *Heart of Matter*, 51.
41 See for example Lukas and Lukas, *Teilhard*, 63. Note that Teilhard himself does not characterize the dualism of matter and spirit as Thomistic or Aristotelian.

movements that saw it as the basis of all evil. John of Damascus went so far as to proclaim, "I shall never cease to venerate that matter through which my salvation was accomplished."[42] Certainly, despite Christianity's theological commitment to matter, dualism was a periodic and recurring problem in actual Christian belief, practice, and language. Yet these moments of departure hardly represent the story of Christian theology as a whole, nor did Teilhard see the Christian tradition as inherently dualistic.

Rather, the threat of dualism became exacerbated by the historical developments that brought about the end of Catholicism's dominance over Western thought: the Reformation, Enlightenment, and the rise of modernity. Martin Luther's (1483–1546) sharp and unrelenting distinction between faith and works, spirit and flesh, was but an early moment in a long line of dualities that fueled modern curiosity: faith versus reason, nature versus freedom, church versus state, individual versus community, the real versus the Ideal. Modern thought can be seen as a diverse showcase of philosophical, theological, and political attempts to navigate such divisions. The myriad of ways to conceptualize the relationship between such elements fall roughly into three overarching strategies. Some seek to maintain the mutual autonomy of each element; some establish a decisive hierarchy between them (colonization); while others attempt to reintegrate both elements into a unified whole. Thus John Locke (1632–1704) sought to secure the divide between public truth and private opinion; Immanuel Kant (1724–1804) relativized the meaning of faith by rehashing it in terms of "practical reason"—a lesser, deponent form of reason; and Friedrich Wilhelm Joseph von Schelling (1775–1854) reintegrated freedom and necessity by seeing both as primordial, irreducible elements of God and the world.

Following this pattern, pioneering intellectuals such as René Descartes (1596–1650) bequeathed unto modernity a tendency to view spirit and matter as existing in sharp opposition to one another. This duality gave birth to modern materialism; only because spirit was seen as incongruous and incompatible with matter could it then become excised from the philosophical system. To the extent that discourse about spirit points to some otherworldly, inaccessible reality (e.g., Kant's *noumena*), it has no real meaning within *this* world and its reality. Yet a material world drained of spiritual depth is inherently a world without meaning; it happens by mere chance and never

---

42 Christoph von Schönborn, *God's Human Face: The Christ-Icon*, trans. Lothar Krauth (San Francisco: Ignatius, 1994), 195.

arrives at any deeper fulfillment or truth. Amid the many responses to this challenge, the British Hegelians of the late nineteenth to the early twentieth century—following the example of Hegel himself—sought to reintroduce meaning into the world precisely by seeing the history of matter as the very unfolding of spirit.

This complex background reveals the significance of Teilhard's approach to the problem of matter and spirit. Far from nullifying the difference, Teilhard highlights the inner unity of the two in a way that is both deeply traditional and relatively novel. His language is replete with references to matter and spirit precisely because this distinction becomes for him the basis of an even deeper unity. Just as the unification of humankind does not destroy individuality, the unification of matter and spirit is not based on "nonduality" (e.g., Meister Eckhart's claim that "between the only-begotten Son and the soul there is no difference")[43] but rather on unity-in-difference. Matter and spirit can only be united because they are, in some sense, distinct. In fact, a true and perfect union of matter and spirit must incorporate and elevate the distinctiveness of both elements, without dissolving one into the other.[44]

The traditional, Catholic character of this union lies in its reliance on a radical *a priori*. A Hegelian dialectic, for example, unifies two elements because they are, at their base, essentially incompatible opposites.[45] A traditional Catholic approach, in contrast, sees the ultimate unity of the elements as the restoration and elevation of an original unity that precedes all division. Hence grace and free-will come together precisely because they were never intended to be at odds with one another within God's perfect plan. As Augustine and Aquinas teach it, grace frees free-will from its slavery to original sin. Free-will was created so that humans might choose grace.

In the same way, Teilhard's perspective does not see spirit and matter as intrinsically opposed but rather as mutually open and co-implicated in God's design for creation. They point to one another. Matter is the formal condition of the possibility of spirit, while spirit is the inner, animating dynamism of matter. Spirit and matter are called to become ever more deeply intertwined precisely because they proceed from one and the same divine will. Viewed from the perspective of evolution, spirit and matter are "no longer two things,

---

[43] Meister Eckhart, *Meister Eckhart: Teacher and Preacher*, trans. Bernard McGinn (New York: Paulist Press, 1986), 264 (sermon 10).
[44] Teilhard de Chardin, *Human Energy*, 104.
[45] This will be explained in depth in chapter 3.

but two *states* or two aspects of one and the same cosmic Stuff."[46] Evolution thus becomes visible as the meaningful and irreversible process of "the progressive Spiritualization of Matter." Spirit is not a distinct and incongruous substance over and against matter. It is the very heart of matter itself, the inner orientation of created matter toward self-transcendence.[47]

Teilhard is neither the first nor the last to deal with this problem in this way. The Anglican theologian John Polkinghorne, for example, takes much the same standpoint.[48] Similarly, the Catholic theologian Karl Rahner approaches the unity-in-difference of spirit and matter from a more systematic, Thomistic angle.[49] Though not entirely lacking in controversy, the idea of the inner unity between matter and spirit recurs in Catholic thought precisely because it stems from traditional Catholic commitments. The interrelation of spirit and matter proceeds from a more fundamental, inner, *a priori* unity, which precedes any apparent division.

Teilhard's positive, unitive approach also colors his understanding of evolution as a whole. The modern propensity for seeing evolution as a "struggle for existence" or "survival of the fittest" goes hand in hand with the tendency to see philosophy as an intellectual battle between dualities such as nature and freedom, matter and spirit. In contrast, Teilhard, like Bergson before him, sees evolution not as the struggle for dominance but rather as the creative unfolding of a fundamental unity underlying all of creation.[50] He believes that the trajectory of evolution leads away from violence, not into it. Accordingly, he understands evolution in its essence not as a violent struggle but rather as a kind of "cosmic birth." Development is painful, difficult, but ultimately good. Through spiritual evolution, the cosmos struggles to overcome enmity and unite its members toward a common goal.[51]

---

46 Teilhard de Chardin, *Heart of Matter*, 26; see also *Future of Man*, 14.
47 See Teilhard de Chardin, *Christianity and Evolution*, 96.
48 J.C. Polkinghorne, *Scientists as Theologians: A Comparison of the Writings of Ian Barbour, Arthur Peacocke and John Polkinghorne* (London: SPCK, 1996), 29.
49 See chapter 4.
50 Bergson, *Creative Evolution*, 116; Teilhard de Chardin, *Building*, 54: Some commentators make a distinction here between Teilhard's evolution as creative unification and Bergson's evolution toward divergence. Ian G. Barbour, "Teilhard's Process Metaphysics," *The Journal of Religion* 49, no. 2 (1969): 148; Arthur Peacocke, *Evolution: The Disguised Friend of Faith?: Selected Essays* (Philadelphia: Templeton Foundation, 2004), 27. Yet this difference is perhaps not as substantial as it may at first appear. In particular, Teilhard does not reduce all evolution to unification but rather sees unification as evolution's final and definitive stage and the primary occupation of the noosphere.
51 Teilhard de Chardin, *Building*, 57, 67.

Seeds of Controversy: Optimism, Evil, and Metaphysics

It almost goes without saying that Teilhard de Chardin's theology is colorfully optimistic. To some extent this makes him emblematic of the early twentieth century, which nurtured many exciting visions of a future without war, without suffering, and without divisions. Nevertheless, Teilhard's optimism almost stands apart. While many early optimists found their hopes tempered by the brutal reality of two world wars, the same cannot be said for Teilhard. If anything, these tragedies led him to cling even more tightly to his confidence in a brighter future for all.

In fact, it was principally this extreme optimism—rather than his acceptance of evolution as such—that made Teilhard's theology controversial within Catholic circles. The first instance of major controversy in his career came to a head in 1925 over an earlier essay on interpreting the doctrine of original sin in light of human evolution, in which he denied the possibility of a historical Adam.[52] Censored by his Jesuit superiors, Teilhard was forbidden to publish his theological musings during his lifetime. Such a restriction did little to silence him. Teilhard continued over the next thirty years to share his ideas in lectures, letters, essays, and even privately circulated books. Following his death in 1955, no longer requiring ecclesiastical approval, his admirers eagerly published his manuscripts for popular consumption.

While the censorship of Teilhard de Chardin was in many ways indicative of the anti-modernist forces at work within the Catholic Church at the time, it would be a misrepresentation to see it merely as an ignorant or reactionary reflex spurred by a lingering fear of new ideas. In fact, Teilhard's evolutionary science was not of itself at issue. He was never condemned for being a scientist. Rather, wrapped up into a superstructure of moral and cosmic meaning, Teilhard's views on evolution had thoroughly crossed the boundary between empirical science and theological speculation. It was this—Teilhard's *theology*—that came under scrutiny.

Teilhard's bubbling optimism was but a key symptom of the potential fault lines running through his theological foundation. Time and again, Teilhard's enthusiasm for a purposeful future runs alongside a marked tendency to downplay the gravity of sin and suffering in the world. In light of the overwhelmingly upward current of evolution, evil appears as nothing more than inertia, lethargy, waste, or a failure to go with the flow. Human violence is

---

[52] This essay was posthumously published; Teilhard de Chardin, *Christianity and Evolution*, 45–55.

an obsolete remnant of our evolutionary past, a representation of our animal drive toward divergence, which is annulled by the rising spiritual current of convergence.[53] In light of this, Teilhard proclaims:

> Evil, in all its forms—injustice, inequality, suffering, death itself—ceases theoretically to be outrageous from the moment when, *Evolution becoming a Genesis*, the immense travail of the world displays itself as the inevitable reverse side—or better, the condition—or better still, the price—of an immense triumph.[54]

In short, evil is the mere incidental underside of the overarching, positive narrative of evolutionary progress. This vision of history's "triumph" no doubt comforted Teilhard by causing the great injustices of the world to lose their magnitude. "The greatest suffering you can think of," he writes, "will disappear, or even dissolve in a kind of pleasure, provided you can discover a correlatively proportionate achievement of which it has been the price."[55]

Ironically, Teilhard penned the above words of triumph in 1942 during the Nazi occupation of his homeland while exiled in faraway Beijing. His optimism was no less stark in 1945, when he described World War II as "a crisis of birth, almost disproportionately small in relation to the vastness of what it is destined to bring forth."[56] Teilhard believed that, notwithstanding the freedom of particular individuals to choose evil, good "must of necessity eventually prevail on earth."[57] The appearance of humanity and the consequent advent of the noosphere marks a definitive turning point in evolution. The future is not to be defined by our animal tendencies toward conflict but rather by the overwhelming impulse of this new stage of evolution toward cosmic unification—or, in Teilhard's view, toward peace. Even the grim horror of the atomic bomb could not damage Teilhard's confidence—far from it. Instead, Teilhard saw the development of nuclear weapons as further confirmation of humanity's impending progress. "To me it seems," he wrote,

---

53 Teilhard de Chardin, *Future of Man*, 145.
54 Teilhard de Chardin, *Future of Man*, 82.
55 Teilhard de Chardin, *Human Energy*, 88.
56 Teilhard de Chardin, *Future of Man*, 110.
57 Teilhard de Chardin, *Future of Man*, 147. Teilhard alternates between confident statements of certain victory and more reserved statements recognizing the contribution of human freedom. Because of this, readers differ on their assessment of the extent to which Teilhard believed that the triumph of good was inevitable. See for example Henri de Lubac, *The Religion of Teilhard de Chardin* (New York: Desclee, 1967), 43.

"that thanks to the atomic bomb it is war, not mankind, that is destined to be eliminated."[58] The devastating power of such a weapon, he reasons, will suffice to ward people off from ever again using it. At the same time, Teilhard believed with the utmost earnestness that, someday soon, scientific curiosity would so overwhelm the human heart and mind as to put an end, once and for all, to our lust for military conquest. Once love of learning overtook the human race, we would set aside our grievances and strive shoulder to shoulder to uncover and admire the mysteries of the universe.

Clearly, two world wars were not enough to break Teilhard's confidence. These global conflagrations instead led him to take refuge in his unshakable evolutionary faith in human progress. Built upon humanity's "natural religious energy," this "faith in man" both stood alongside and flowed upward into Teilhard's "supernatural" faith in Jesus Christ.[59] Viewing the world and its history through this lens allowed Teilhard to see even the most bitter moments of suffering as contributing, however mysteriously, toward the ultimate triumph of the divine plan. He needed this. Having served as a stretcher-bearer and chaplain in the First World War—a conflict that claimed the lives of two of his brothers—he relied upon his faith in progress in order to find meaning despite the horror and irrationality of it all.[60] This same faith carried him through the war's sequel.

In other words, Teilhard's optimism provided him with comfort. It served to diffuse the problem of evil. It was, in other words, a theodicy.[61] Suffering and death are natural and intrinsic elements of biological evolution. Accordingly, for Teilhard, they must also be necessary and unavoidable concomitants of God's benevolent plan for the world.[62] In this way, Teilhard makes little distinction between natural evils (earthquakes, disease, etc.) and moral evils (human violence). As "an inevitable byproduct" of the process of unification, evil must come about "*as a matter of statistical necessity.*"[63] For Teilhard, this

---

58 Teilhard de Chardin, *Future of Man*, 140, 148.
59 Pierre Teilhard de Chardin, *Toward the Future*, trans. René Hague (San Diego: Harcourt, 2002), 23–30.
60 See Pierre Teilhard de Chardin, *The Making of a Mind: Letters from a Soldier-Priest, 1914–1919*, trans. René Hague, 1st ed. (New York: Harper and Row, 1965), esp. 171.
61 Teilhard de Chardin, *Toward the Future*, 196–97; *Christianity and Evolution*, 196.
62 See Teilhard de Chardin, *Phenomenon of Man*, 309; *Christianity and Evolution*, 82, 225; *Science and Christ*, trans. René Hague (New York: Harper and Row, 1968), 80.
63 Teilhard de Chardin, *Toward the Future*, 198; *Christianity and Evolution*, 196.

is good news. As Christ's death on the cross epitomizes, God can bring a greater good out of even the most horrific evil.[64]

Though laudable for their message of hope, Teilhard's rosy remarks can pose serious theological and pastoral difficulties. To claim that the eventual "triumph" of evolution justifies suffering all too easily brushes aside the plight of those who, caught up in the machinations of progress, pay far more than their share of the "price" of progress. Can the Armenian people see the genocide of 1915–17 as fair payment for the progress of humanity? Would the victims today of China's mass imprisonment, persecution, and cultural genocide of the Uyghurs agree that any progress is being made at all? Will a starving laborer in one of Peru's toxic gold-mining slums, whose nervous system bears the permanent scars of mercury exposure, find anything to celebrate in the supposed "triumph" of evolution?

Christianity often faces the challenge of proclaiming God's triumph over evil without downplaying the real, tangible horror of evil in the world. Simply because Christ redeemed us by undergoing violence, malice, and death does not make these things good in themselves. God brings about a greater good not so much *because of* evil as *despite* it, for evil actively works against the progress of good in the world.

Accordingly, Teilhard's words are at risk of glorifying evil by effacing the real difference between it and the good. Teilhard's economic analogy illustrates this problem well. To characterize evil in terms of "price" implies a rate of exchange. A price is quantitatively convertible with that which it buys. A certain number of Japanese yen can be exchanged for an equivalent number of Australian dollars. Teilhard thus speaks of evil as though it necessarily and almost mechanically converts into the good by means of the natural process of evolution. Evil becomes a sort of evolutionary investment toward the promise of staggering returns. To thus make evil convertible into good both downplays the real danger of evil and dulls the sharp edge of Christianity's condemnation of immorality.

Teilhard's admirers have been deeply divided in regard to such claims. Henri de Lubac, for example, defends Teilhard's orthodoxy with an unwavering confidence in the scientist's genuine appreciation of Catholic tradition. De Lubac sees Teilhard's most controversial comments as mere "isolated

---

64 Teilhard de Chardin, *Human Energy*, 51–52; cf. *Phenomenon of Man*, 311.

aberrations."[65] Teilhard was no willful agitator against Church teachings but rather a victim of his own "clumsy" expression.[66] From de Lubac's standpoint, the bulk of Teilhard's writings, as well as Teilhard's own most basic intentions, remain faithful to the authority of Catholic doctrine.

In contrast, many current Teilhardians favor reading Teilhard in precisely the opposite direction. Noting the consistency of Teilhard's theodicy with his views as a whole, they see his position on evil as central to his understanding of God and the world. Where De Lubac saw a misunderstood pillar of orthodoxy, Ilia Delio sees a great reformer and "new Elijah" who pushes back against the dogmatism of the Church and its reliance on an Augustinian doctrine of original sin.[67]

Delio's progressive reading of Teilhard, which she shares with such theologians as Karl Schmitz-Moormann and Ursula King, highlights Teilhard's approach to evil precisely because it leads him to cross from a metaphysics of the world into a metaphysics of God.[68] The bulk of his work focuses on the former. It proposes a metaphysics of the world by reconceptualizing creaturely existence in terms of the evolutionary process. Past, present, and future become dimensions of the creature's progress toward ultimate fulfillment. In contrast, Teilhard's rare explicit treatments of theodicy shift into clear metaphysical claims about the very nature and operation of God. The existence of evolution, he argues, implies an inner necessity that structures God's creative action. The fact that God has created the world by way of evolution means, he insists, that "God cannot create except evolutively."[69] The reality of evolution tells us something about what God can and cannot do. It provides, in other words, a metaphysical conception of God's own divine nature. For Teilhard, this limitation on God's creative ability actually derives from God's perfection; the self-communication of God to creatures requires a gradual, evolutive process by which creation becomes acclimated to receive its Creator.

---

65 De Lubac, *Religion of Teilhard*, 195. De Lubac may not have been aware of Teilhard's *Christianity and Evolution* when he began to write in Teilhard's defense.

66 De Lubac, 38.

67 Ilia Delio, *The Unbearable Wholeness of Being: God, Evolution, and the Power of Love* (Maryknoll, NY: Orbis, 2013), xvi, 117–17.

68 On Teilhard's metaphysics, see David Grumett, "Metaphysics, Morality, and Politics," in *From Teilhard to Omega: Co-Creating an Unfinished Universe*, ed. Ilia Delio (Maryknoll, NY: Orbis, 2014), 111–26.

69 Teilhard de Chardin, *Christianity and Evolution*, 179; *Toward the Future*, 197.

From this standpoint, Teilhard takes aim at the traditional Catholic doctrine of creation. Traditional metaphysics emphasizes the intrinsic perfection and inner completion of God in contrast to the dependency and nonnecessity of creation. A self-sufficient God creates the world out of an entirely gratuitous love, which desires absolutely nothing in return. But this also means that creatures are fundamentally unnecessary; God did not *have to* create the world. This does not mean that God is indifferent or uncaring. Quite the contrary, it means that God creates the world out a love that is so free as to require nothing in return. God does not create the world for the sake of God's own gain. Creation is pure gift.[70]

Nevertheless, Teilhard finds this traditional view "humiliating" and "offensive."[71] In his view, the nonnecessity of creation undermines his explanation of evil by implying that the world does not serve an ultimate purpose. The "radical uselessness" of creatures provides no satisfying justification for the evolution's requisite price of suffering and pain.[72] Therefore, he argues, we must rethink the dogma concerning God's complete freedom *not* to create the world. In contrast to the traditional association of God with self-sufficient *being*, Teilhard proposes a metaphysics rooted in union and participation. He suggests that just as God necessarily becomes a Trinity, entering into multiplicity in order to realize a profound union, so also does this movement flow outward into the act of creation. God thus creates for the sake of achieving perfect unity with the world. In some sense, this goal must be beneficial not only for us but also for God as well. Once again, Teilhard tends to think in economic terms; he insists that the creation of the world only makes sense if the ultimate unity of Christ and the world will "be worth what it has cost God."[73]

To be clear, Teilhard does not actually say that God *had to* create the world. Rather, in a manner reminiscent of the theology of the Russian theologian

---

70 This doctrine is often criticized today. For a detailed treatment of the ongoing debate and a thorough presentation of the traditional doctrine, see Thomas G. Weinandy, *Does God Suffer?* (Notre Dame, IN: University of Notre Dame, 2000). Contemporary Catholic criticism of the doctrine is deeply indebted to Catherine Mowry LaCugna, *God for Us: The Trinity and Christian Life* (San Francisco: HarperSanFrancisco, 1991). LaCugna relies significantly on the work of Piet Schoonenberg, who was himself indebted to Teilhard.
71 Teilhard de Chardin, *Christianity and Evolution*, 226.
72 Teilhard de Chardin, *Christianity and Evolution*, 225.
73 Teilhard de Chardin, *Science and Christ*, 17.

Sergei Bulgakov (1871–1944), Teilhard wants to suggest that the world is more intimately part of God's own self-fulfillment than the doctrine of the world's nonnecessity would imply. On the one hand, the world is not *strictly* necessary. It is not a sort of mechanical outgrowth of divinity, something God must simply accept as required by God's own nature. On the other hand, neither is the world so contingent as to make God inherently indifferent to its existence. The trick is to have one's cake and eat it too: to maintain God's independence and transcendence without making the world merely accidental or utterly meaningless by comparison. For his part, Bulgakov works out this paradox by arguing, "If God created the world, this means that he could not have refrained from creating it."[74] God was unable *not* to create the world. Yet this is not because God is compelled to create by any external law but rather because only by creating the world could God live in accord with God's own inner essence. Despite Bulgakov's careful arguments, one can see his 1935 condemnation by both the Russian Orthodox Patriarchate and the Russian Church Outside of Russia as indicating that his views failed to gain significant support.

Teilhard's attempt to walk this same tightrope remains ambiguous and underdeveloped by comparison. He frequently backs away from his most daring claims with contradictory professions of orthodoxy. With one breath he argues that God can no longer be considered "structurally independent" from creation, but with the next he insists that God does not become "lost" in the ongoing process of creation.[75] In one instant he maintains that an orderly world cannot be created without intervening disorder (i.e., suffering and sin), but in the next he adds that nothing "impairs the dignity or limits the omnipotence of the Creator."[76] Thus, while in Delio's interpretation the proper emphasis falls upon Teilhard's rejection of creation's nonnecessity, David Grumett sees Teilhard as affirming—in line with Catholic tradition—the paradoxical fullness of God's immanence *and* transcendence.[77]

As we shall see next chapter, Delio's progressive interpretation puts Teilhard in good company with Alfred North Whitehead's process theology. Those contemporary theologians who favor Teilhard's theodicy and its attendant claims about creation and the divine nature are often influenced by

---

74 Sergius Bulgakov, *The Bride of the Lamb*, trans. Boris Jakim (Grand Rapids, MI: Eerdmans, 2002), 31.
75 Teilhard de Chardin, *Christianity and Evolution*, 239.
76 Teilhard de Chardin, *Toward the Future*, 197.
77 Grumett, *Teilhard de Chardin*, 221–24.

process theology, which makes such claims in an even more vivid and less-reserved form. For such theologians, theodicy typically stands not as a mere product of the evolutionary worldview but rather as a fundamental starting point of evolutionary theology as such. Theodicy becomes *the* fundamental question. Moreover, if Teilhard's approach suggests a reconsideration of the nature and freedom of God, process theology will *demand* it instead. Where Teilhard's evolutionary theology remains largely rooted in a metaphysics of the world, process theology is every bit as much rooted in a corresponding metaphysics of God.

Christocentricity and Personalism

Nevertheless, Teilhard's approach also stands out over and against process theology because of his core insistence upon the centrality of Jesus Christ. Teilhard sees this as *the* primary doctrine of Christian faith,[78] and his zeal for Christ permeates his thought. As he understands it, evolution is the process by which the world enters into an ever deeper encounter with its Creator, and Christ *is* the very embodiment of this encounter.[79] Drawing on the biblical language of Christ as *alpha* and *omega* (Rev 1:8, 21:6, 22:13),[80] the beginning and the end, Teilhard sees Christ as the originator, focal point, and goal of all evolution—planetary, geological, biological, as well as spiritual.[81] Evolution is no mere general orientation of matter toward spirit. It is ultimately a specific trajectory toward a "well-defined supra-personal focal point," the incarnate Jesus Christ, in whom the entire evolutionary narrative is definitively and irreversibly revealed as the movement of biosphere toward noosphere.[82] Teilhard's theology is *Christocentric* to the extreme.

Though rooted in his Catholic spirituality and upbringing, Teilhard's Christocentrism also carries a distinctly universalist bent. In his view, Christ is already present in and to the world prior to and beyond the ministry of the Catholic Church. He does believe that the Catholic Church has a unique significance for revealing Christ to the world. Catholicism is not simply one religion among

---

78 Teilhard de Chardin, *Christianity and Evolution*, 190; *Making of a Mind*, 173.
79 Teilhard de Chardin, *Future of Man*, 307–307.
80 A and Ω, the first and last letters of the Greek alphabet.
81 Teilhard de Chardin, *Christianity and Evolution*, 180.
82 See Teilhard de Chardin, *Heart of Matter*, 44.

others. However, Teilhard also firmly believes that human religiosity—even pagan religiosity—is rooted in a natural spiritual impulse or "energy" toward Christ. In fact, for Teilhard, even science is in some sense a spiritual quest for the face of Christ written upon the evolving universe. In this way, Teilhard's Christocentrism is not reductively Catholic. Rather, in his view, the doctrine of Christ is the authentic crystallization of an intrinsic and authentically human impulse.

Nevertheless, the contours of Teilhard's understanding of Christ are still largely shaped by his Catholic spirituality. More specifically, Teilhard sees the mystical tradition as encapsulating a more pure and original understanding of Christ, to which doctrine must attend.[83] Teilhard does not necessarily reject the Church's dogmatic understanding of Christ, but he does believe that doctrine needs to be conceptualized anew in light of both the mystical tradition and the modern evolutionary perspective. Such an approach does not disregard Scripture. For Teilhard, a turn toward this mystical, universal Christ is an authentic return to the Christ of the Bible. Even so, this illustrates how Teilhard's reading of the biblical Christ focuses less on his human deeds and more on the transcendent Christ of Paul and John, the Christ whose action in the world precedes his human life (e.g., Col 1:15-20; Phil 2:6-11; John 17:1-5).[84]

Yet in view of Catholic doctrine, Christ's transcendence goes hand in hand with his intimacy with creation. Teilhard recognizes how the dogma of the *hypostatic union*, which affirms complete, unmitigated, and unadulterated humanity and divinity of Jesus Christ, functions as the definitive confirmation of the fundamental compatibility of creation with its Creator.[85] The real, concrete God-man Jesus Christ, irreducible to a mere myth, idea, symbol, or concept, enlivens the history of this evolving world precisely by taking its material existence as his own.[86] Through the incarnation, Christ reveals the inner meaning and purposefulness of the world even as the world becomes itself the revelation of Christ. For Teilhard, the evolving character of this world is integral to its revelatory function. "By disclosing a world-peak," he writes, "evolution makes Christ possible, just as Christ, by giving meaning and direction to the world, makes evolution possible."[87] In other words, because evolution is (in Teilhard's view) the progressive spiritual unification

---

83 Teilhard de Chardin, *Science and Christ*, 15–16.
84 See Teilhard de Chardin, *Making of a Mind*, 93.
85 See Teilhard de Chardin, *Christianity and Evolution*, 69–71.
86 Teilhard de Chardin, *Christianity and Evolution*, 88.
87 Teilhard de Chardin, *Christianity and Evolution*, 128.

of the world, it is essentially the revelation of Christ as "all in all" (Col 3:11).[88] Evolution unveils the inner unity of God and the world in Jesus Christ, which only deepens as time moves forward.

Importantly, Christ's central significance ties in closely with Teilhard's essential *personalism*. Teilhard's universal, cosmic view of evolution, which envisions the world as one living organism, does not overlook the value and significance of the individual person as one incidental atom of this overarching process. Rather, the meaning of the individual is deepened by the authentic unity of the cosmic multitude.[89]

Christ seals and confirms this emphasis on the person precisely because Christ is himself a person, a man, and not a mere symbol or concept. This transcendent Christ who is *alpha* and *omega*—the origin and goal of the evolutionary process—is the very same Jewish man who dwelled on earth at a definite time and place. Christ is fully universal and fully personal at the same time. In fact, it is because Christ is a concrete, individual person that the drama of evolution can never be reduced to a mere general theory that operates solely on the level of impersonal, collective realities such as populations, species, societies, or worlds. The development of the world is intimately linked to the dignity and freedom of the individual, who is an indispensable organ of the emergent noosphere.

Such personalism further underlines the intimacy of God's relationship with the world. For many, Darwinian evolution points to a God who remains distant and impersonal, uninvolved in the ordinary processes of the world. In essence, however, Teilhard's view requires the exact opposite. Only a personal God, a God whom one can address with prayer and sighs, can be the author of such a dynamic world as Teilhard envisions. As Henri de Lubac explains,

> For Teilhard, everything rests on the primacy of the person. The concrete Presence at the heart of the Universe . . . —super-personal, i.e., ultra-personal—of a loving and provident God, of a God who can reveal himself, and has in fact revealed himself . . . —this was for Teilhard the supreme truth.[90]

---

88 Teilhard de Chardin, *Science and Christ*, 54, 164–67.
89 Teilhard de Chardin, *Future of Man*, 303 n. 1; *Toward the Future*, 36, 150–51; *Christianity and Evolution*, 129, 171; *Human Energy*, 61–67, 142–44. See also Henri de Lubac, *Teilhard de Chardin: The Man and His Meaning*, 1st ed. (New York: Hawthorn Books, 1965), 21–27.
90 De Lubac, *Teilhard de Chardin*, 23.

As evolution transitions from biosphere to noosphere, it ceases to be solely a material process and becomes all the more spiritual. This effectively awakens and radicalizes the personhood of the individual; and yet, this individuation is far from individualistic or selfish. Rather, the person is drawn into a universal process of "amorization"—of conversion to love (in French, *amour*)—whereby a universal love of all, for all, and by all becomes the definitive form and final realization of this evolving world.[91] The gradual deepening of the individual operates precisely so that one may enter into a free and liberating relationship with Christ in the universe and the universe in Christ. Thus, while Teilhard envisions the ongoing trajectory of human evolution as "convergent"—tending toward a deep unification of hearts and minds—this spiritual unification does not annul our unique differences, making us nothing more than mere instances of some collective human consciousness. Rather, it elevates personal differences, universalizing the Catholic idea of unity-in-difference whereby the distinctiveness of the individual actually serves to concretize the deeper unity of humanity as a whole. Such a unity can only be ratified in the appearance of a distinct individual, Jesus Christ, who at one and the same time is in himself the progressive unification of the world.

TEILHARD'S EVOLVING SIGNIFICANCE

These key themes from Teilhard's theology continue to exert a significant influence on contemporary evolutionary theology. Nevertheless, there is more than one way to tell the history of evolutionary theology. After all, evolutionism is in essence a mode of conceptualizing history. It is one expression of the modern tendency to view intellectual shifts and trends in terms of a web of influences and interconnected ideas that form a continuous history of development. Such narratives acquire particular force when, for the sake of tying the strings of fate into one tidy knot, they point to some exceptional genius as *the* critical nexus of innovation: a Galileo, a Newton, or an Einstein, one who towers over all others and establishes history in terms such as "before Napoleon" and "after Napoleon." Even today Darwin is often credited as "the father of evolution" in a way that completely forgets Wallace's contribution. Teilhard could easily serve as such a figurehead. A vast number

---

91 Teilhard de Chardin, *Heart of Matter*, 50–52.

of evolutionary theologians draw inspiration from Teilhard to a greater or lesser extent.

As we have seen, however, Teilhard is just as much the product of a tradition. His genuine creativity is intimately linked to the way in which he serves also as a vital conduit for the development of earlier ideas. He did not invent organicism, Christocentrism, or even personalism, and yet he is famous for deploying these elements together into a fluid, dynamic, and inspiring vision of God's intimate relationship with the evolving world. The wellsprings of evolutionary theology did not dry up with Teilhard, however. Accordingly, some figures such as the Anglican priest, physicist, and theologian John Polkinghorne share much in common with Teilhard without being reducible to his disciples. Polkinghorne's theological project resembles that of Teilhard not only because they share many of the same conceptual elements but also because both men are scientists who are convinced that science and theology must enter into a mutually constructive dialogue.

Most importantly, many of the same influences that gave birth to Teilhard's theology are responsible for the construction of another, distinct and influential tradition—another founding figure who stands just as tall, if not taller. Though its influence peaked more slowly, Alfred North Whitehead's process theology grounds its own distinct brand of evolutionary theology.

As we shall see, the ongoing story of evolutionary theology can often be read in terms of these two influences: Teilhard and Whitehead. There are those who follow one, those who follow the other, and those who pursue a mixture of both (with particular emphasis on one source or the other). Only a small minority stand altogether outside of this spectrum. Yet even unique figures such as John Polkinghorne cannot altogether escape the Teilhard-Whitehead dynamic, since Polkinghorne will often be interpreted by his readers in either a Teilhardian or a Whiteheadian manner. What makes these two so influential, and how does a Whiteheadian interpretation shape one's reading of evolutionary theology? We turn now to discover these answers.

CHAPTER THREE

Process Theology and Hegelian Dialectic
Evolution as Co-creation

As the case of Teilhard de Chardin has illustrated, evolutionary theology arose among thinkers who believed not merely that the science of evolution could inform our theological understanding but also that evolution itself is in some sense *revelatory*. To be revelatory is more than to be informative. Any idea, concept, theory, or fact might be informative. A detailed study of seventeenth-century German pietism, for example, would provide important data for understanding its influence on subsequent Lutheran theology and practice. By comparison, to see evolution as truly revelatory is to assert not that it provides useful information but rather that it uncovers some deeper, more fundamental, even transcendent *meaning*.

Teilhard saw evolution as revelatory precisely because the very process of evolution—not merely the theory of evolution, but evolution itself—moves toward the ultimate fulfillment of the world. Evolution is for him the story of the rise of consciousness toward universal love. The world writes its own story of growth as it progresses, and in this sense evolution appears also as the process of the world's definitive self-revelation. It is the movement of historical reality toward the achievement of its own defining ideal; it is the contingent but creative process by which the world comes to be what it most authentically is.

But why stop there? What if evolution reveals something more, something beyond our usual conception of the world? Teilhard's periodic forays into the theology of God, which tested or even traversed the bounds of his Catholic orthodoxy, betrayed the tenuousness of the boundary between an evolutionary theology of the world and an evolutionary theology of *God*. This tenuousness was itself the product of evolution's elevation from a biological theory to a philosophical principle. A pure Darwinian perspective can see the transmutation of species as a mere fact-based description of what tends to happen in the realm of biological nature. Philosophers and theologians of evolution, however, tend to assume in principle that evolution has some fundamental significance; if not directly instituted by the divine, then at least it must be a structural principle within reality as such. Hence many have seen evolution as occurring not merely among organisms but even among planets, rock formations, continents, cultures, religions, governments, ideas, and so forth—so why not God as well?

The development of an evolutionary theology of God was thus all but inevitable. It reached an important apex in the process theology of Alfred North Whitehead,[1] which functions as a second pillar of contemporary evolutionary theology. Even in the midst of rapidly growing theological diversity in the twenty-first century, process theology often forms a critical common ground underlying many key concepts, tenets, and methodological decisions.

Accordingly, any systematic overview of contemporary evolutionary theology must explore Whitehead's continuing influence and significance. A full examination of his philosophy would require a dedicated undertaking of its own and thus exceeds the limits of the present study. In this short space, the best we can hope to accomplish is to highlight the intellectual heritage and key contributions of Whitehead alongside some important applications of these ideas. This will require, moreover, a brief examination of the relation between Whitehead's philosophy and that of Georg Wilhelm Friedrich Hegel.

This brief overview will make it clear that process theology's principal contribution to evolutionary theology is its expansion of the evolutionary paradigm beyond the realm of creation alone, so that even God in God's own self becomes illuminated by a metanarrative of progress and gradual development. More and more, many evolutionary theologians have come to see

---

[1] We follow here the now common practice of reserving the term "process theology" to refer to the specific school of thought directly influenced by Whitehead. Nevertheless, as Ewert Cousins points out, this term has also been used in many other ways, including a more general way that would include Teilhard as well. *Process Theology: Basic Writings* (New York: Newman, 1971), 8–9. Note also that while some restrict the term to Whitehead's successors—since Whitehead's own work is more properly named "process philosophy"—the fact is that Whitehead's philosophy is already through and through theological in substance.

evolution not only as the history of the world but also as the very history of God in and through the world—a continual unfolding of divine self-creation. In view of this dynamic, we creatures are by extension co-creators who form the world and its salvation even from within.

THE MOVEMENT FROM HEGEL TO WHITEHEAD

It is no exaggeration to say that Georg Wilhelm Friedrich Hegel (1770-1831) has exerted a monumental influence upon subsequent thinkers, both in his day and in ours. Gary J. Dorrien avers that "every post-something theory or critical approach traces in some way to Hegel."[2]

Yet like most great innovators, Hegel's originality does not negate the fact that he is also a conduit and refractor of earlier traditions and parallel influences. Formed by Immanuel Kant, inspired by Johann Gottlieb Fichte (1762-1814), and in competition with F. W. J. Schelling, Hegel took up the developmental perspective pioneered by figures such as Leibniz and intensified it until it reached a boiling point. From there, this developmental perspective came to color virtually everything that followed after. It fueled the evolutionisms of the nineteenth century before and after Darwin. Hegel's ideas filled the very air of the nineteenth century. Thus, Darwin had to present his theory of natural selection within a world and a language irreparably shaped by Hegel. Friedrich Nietzsche went so far as to say that "without Hegel there could be no Darwin."[3]

Perhaps even more poignantly, without Hegel there could be no Whitehead. While Whitehead formulates his own distinct language and expression, the major brushstrokes of a Hegelian metaphysics are all visible. At times, Hegel's ideas are so recognizable within Whitehead's philosophy that Whitehead almost seems to have merely translated them for a newer context. In the last analysis, process theology is more distinctly Hegelian than Darwinian.[4]

---

[2] Gary J. Dorrien, *In a Post-Hegelian Spirit: Philosophical Theology as Idealistic Discontent* (Waco, TX: Baylor University Press, 2020), x.

[3] Friedrich Wilhelm Nietzsche, *The Gay Science: With a Prelude in German Rhymes and an Appendix of Songs*, ed. Bernard Williams, trans. Josefine Nauckhoff and Adrian Del Caro (Cambridge: Cambridge University Press, 2001), 218 (§357).

[4] On Whitehead's understanding of evolution, see George R. Lucas Jr., "Evolutionist Theories and Whitehead's Philosophy," *Process Studies* 14 (December 1, 1985): 287-300.

It is striking, in view of this heritage, how little explicit reference Whitehead actually makes to Hegel himself. Not only does he seldom mention Hegel, but he even admits to having read almost nothing of Hegel's writings.[5] Some interpreters take this relative lack of a direct connection as proof of Whitehead's originality. A few of Whitehead's most dedicated disciples even balk at the very suggestion that Whitehead owes anything to Hegel. As Robert Ellis quips, such a romantic view of Whitehead's originality gives "the impression that Whitehead is a sort of Melchizedek of twentieth-century philosophy—arriving on the scene to bless Abrahams and slay other warriors—without ancestry and any obvious philosophical lineage."[6]

Such passionate insistence on Whitehead's uniqueness clearly owes more to emotional commitments or philosophical party politics than to any rigorous comparison of texts, sources, and themes. In some cases, it stems also from an inadequate understanding of Hegel's views.[7] More importantly, it runs ironically counter to Whitehead's own framework, which sees thought as developing historically through the interaction of a variety of influences. In reality, even though Whitehead did not read Hegel at length, he received Hegel's influence through secondary sources. In particular, as we shall discuss, Whitehead was influenced by the British Hegelianism of F. H. Bradley as well as by its refraction in the work of Samuel Alexander.

Perhaps George R. Lucas Jr. expresses it best when he avers that "in spite of certain differences, Whitehead and Hegel represent two schools within a single tradition of process philosophy."[8] The statement that Whitehead is rooted in Hegel need not be taken as undermining the significance of Whitehead himself. Whitehead certainly is not reducible to Hegel. Nevertheless, the

---

5 Robert Ellis, "From Hegel to Whitehead," *The Journal of Religion* 61, no. 4 (1981): 404.

6 Ellis, 403. Cf. Victor Lowe, *Understanding Whitehead* (Baltimore: Johns Hopkins Press, 1962), 254–57; Charles Hartshorne, "Introduction: The Development of Process Philosophy," in *Philosophers of Process*, ed. Douglas Browning (New York: Random House, 1965), x.

7 See for example Gregory Vlastos, "Organic Categories in Whitehead," *The Journal of Philosophy* 34, no. 10 (1937): 253–262. For a critique of Vlastos, see George L. Kline, "Concept and Concrescence: An Essay in Hegelian-Whiteheadian Ontology," in *Hegel and Whitehead: Contemporary Perspectives on Systematic Philosophy*, ed. George R. Lucas Jr. (Albany, NY: State University of New York Press, 1986), 141–42.

8 George R. Lucas Jr., *Two Views of Freedom in Process Thought: A Study of Hegel and Whitehead* (Missoula, MT: Scholars, 1979), 8.

material connections between Whitehead and Hegel are substantial. Not only do they share many of the same concerns and insights, not only do they both dramatically rethink God in much the same way, but they also evoke many of the same critiques from other forms of Christian theology and philosophy. While many interpreters highlight an assortment of differences that seem to set the two at odds with one another, these remain far less substantial than the innumerable similarities couched in mere terminological idiosyncrasies.

Consequently, some engagement with Hegel is extremely important for drawing out the broader significance of Whitehead's influence on contemporary evolutionary theology, including how it stands relative to other views. In sum, a fuller understanding of Hegel provides a better understanding of Whitehead.

Hegel's Dialectical Logic

A single structured and comprehensive logic underlies Hegel and Whitehead alike. This logic forms their real common ground. It deeply informs their concerns, convictions, and conclusions. This logic is evolutionary in character. Because of this, both Whitehead and Hegel understand the world as essentially dynamic and progressively developing toward a fuller realization of its inner meaning and significance. Moreover, this logic is so fundamental that, for Hegel and Whitehead alike, it necessarily applies just as well to God. In their view, God radically and personally embodies the logic of the evolutionary process.

To better understand this, we have to see how such a logic operates and how, more specifically, these thinkers can move from logic to ontology. In other words, how do we get from a framework for conceptual thought to a metaphysical doctrine about the meaning, character, and movement of being?

In ordinary use, logic provides a formal structure to how we think. It allows us to analyze and interpret events and make predictions based upon these results. As a *formal* element of thought, however, logic does not strictly produce this or that content, this or that truth, and so a wide variety of perspectives can follow the same constitutive logic. Two different crime scene investigators report to the scene of a brazen bakery burglary. The first, reasoning that the culprit must have had a close association with the bakery, fingers the employee on duty, who had the opportunity and means. The second actually uses the same logic—the culprit must be associated with the bakery—but draws a different conclusion. It must be the owner's son, who had intimate knowledge of the crime scene, employee schedules, and security loopholes. Even when deployed well, logic can only point in a useful direction by validating a possible conclusion as logical or illogical. Ordinary logic cannot force a conclusion. Much less can it make a binding claim about the nature of God.

Hegel's view of logic differs here substantially. He insists that genuine logic cannot be merely formal but must determine its content as well. It is never only a conceptual tool that I, as the subject, bring to the study of philosophy. Rather, this logic is ultimately identical with its content; to grasp this logic in the fullest sense is at one and the same time to know truth itself. This logic does not function, however, to reveal incidental truths such as who took the cake. Rather, it reveals "the realm of pure thought. This realm is truth as it is without veil and in its own absolute nature. It can therefore be said that this content is the exposition of God as he is in his eternal essence before the creation of nature and a finite mind."[9] Yet Hegel's logic does not stop at a knowledge of God in the abstract. It is rather the explanation of God as dynamically coming to be through the world and history.

If compressing the truth about God and the world into a single system of logic sounds excessively bold, for Hegel this boldness is precisely what constitutes its validity. Aristotle had defined intellectual disciplines (*epistēmē* in Greek) in terms of their self-contained completeness. Physics was an *epistēmē* because it encapsulated one whole realm of thought and could not simply be broken up into or replaced by other disciplines. Hegel takes this even further: *only that which is self-contained and complete can be true.* There is no partial truth. Any attempt to know just this or that aspect about God, the world, or humanity necessarily becomes lost in mere fragments and incidental details. Accordingly, Hegel's logic is preeminently true precisely because, as that which transcends history, it is able to encapsulate *all* of reality into itself—even its own negation. This is its central principle: that which is genuinely true must contain its own negation.[10] If truth must be absolutely comprehensive, then it must contain even that which stands in stark opposition to itself. The truth about being must contain nonbeing; the meaning of the good must be shaped

---

9 Georg Wilhelm Friedrich Hegel, *Hegel's Science of Logic*, trans. Arnold V. Miller (Amherst, NY: Humanity Books, 1999), §53.

10 Hegel, §62.

by that of evil. By facing its own negation and transcending it, truth becomes something more: the unity of itself and its opposite.

Such an all-unifying logic responds to a distinctively modern tendency toward duality. Beginning roughly in the seventeenth century, various philosophers sought to understand reality in terms of fundamental distinctions between elements such as faith and reason, church and state, private and public, individual and community, nature and freedom, subject and object, God and the world, and more. Often taking the difference between such dualities to be axiomatic, modern philosophers proposed various ways of understanding the interrelation between these different elements. Do faith and reason operate within their own, nonoverlapping spheres? Is one better than the other? Can each contribute something positive to the other? Or are they utterly irreconcilable?

Even when philosophers attempted to form some positive, peaceful relationship between two such elements, the result often failed to provide any real, inner, and definitive unity. For example, some attempts at reconciliation subsumed one element under the other. Kant sought to validate both faith and reason, but he merely reduced faith to a lesser, defective form of reason, which he named "practical reason."[11] This allowed him to see the two as mutually supportive, but only at the expense of faith's distinct validity and autonomy. In the final analysis, "pure reason" and "practical reason" did not stand on equal footing. Kant's very terminology implied that one was technically better than the other, just as premium gasoline is technically better than regular unleaded.

If Kant failed by making faith and reason too much the same, other attempts to reconcile such dualities erred by failing to move beyond apparent differences. Hegel sees Fichte's attempt to bridge the gap between subject and object in this light. Think about this sentence: "She reads a book." The subject (she) freely acts upon the object (the book), and this very act proves her own subjectivity—her freedom over and against the object. The object for its part passively receives this action without having any choice in the matter, and yet ironically it becomes the occasion for the subject to prove its subjectivity. The object provides a *something* upon which the subject may act. Thus, are subject and object really altogether independent? At least in Hegel's

---

11 Immanuel Kant, "Lectures on the Philosophical Doctrine of Religion," in *Religion and Rational Theology*, ed. George Di Giovanni, trans. Allen W. Wood (Cambridge: Cambridge University Press, 1996), 406–409, 415.

view, Fichte was able to show the dependence of objective reality upon the subject, but not the dependence of subjectivity upon objectivity. Despite his attempt to unify the two, subject remained more fundamental. Objective reality—the world "out there"—served only as an external and secondary space for the subject's self-manifestation.

Hegel therefore believes that the task of true philosophy is to press even further toward the reconciliation of such diametrically opposed elements. Not only does objective reality depend upon the subject, but subject depends upon objectivity.[12] The subject realizes itself as subject only by revealing itself, manifesting itself in and through the objective world. It must lose itself in order to find itself, for the truly self-conscious self is that which *necessarily* empties itself into the *other* and rediscovers itself as that very process of self-emptying and self-discovery. In this way, Hegel argues, irreconcilable difference becomes the basis of ultimate reconciliation. Plurality becomes the foundation of true unification.

Hegel terms his all-encompassing logic of reconciliation *dialectic*. This word has a long history. Even today it does not maintain a consistent meaning among philosophers and theologians. For Plato, "dialectic" simply referred to the conversational method of Socrates. Socrates used this discursive method to unveil misconceptions and the illusory character of sophistic syllogisms. In view of this, Kant repurposed the term for his "transcendental dialectic," a subsection of his system of logic that aimed at unveiling the illusory character of certain metaphysical arguments.[13] Though Kant would have dismissed Hegel's dialectic for transgressing reason's limits, Hegel actually praises Kant's subversion of dialectic for one key reason: Kant insisted that such metaphysical illusion and logical contradiction were necessary and unavoidable concomitants of reason itself.[14] In other words, as Hegel sees it, logic goes hand in hand with illogic, and reality goes hand in hand with illusion.

Hegel thus subverts Kant's subversion. He takes this struggle of opposites as the real content of dialectical logic. Such dialectic is not merely one aspect of logic but rather logic's own inner flow. It is reason's propensity for

---

12 Georg Wilhelm Friedrich Hegel, *Hegel's Philosophy of Mind*, trans. William Wallace (Oxford: Clarendon, 1894), 7; *Phenomenology of Spirit*, trans. Arnold V. Miller (Oxford: Clarendon, 1977), §804.

13 Immanuel Kant, *Critique of Pure Reason*, trans. J.M.D. Meiklejohn (London: George Bell and Sons, 1887), 211.

14 Hegel, *Science of Logic*, §68.

transcending all limits and producing truth by engaging and subverting every opposition, even the opposition between the logical and the illogical. By taking opposition as the inner foundation of true unity, Hegel's dialectic reintegrates the dualities of modern thought—faith and reason, individual and community, and so forth—in such a way that their profound oneness becomes rooted at one and the same time in their real opposition to one another.

To put it simply, Hegel's dialectic involves two elements, a positive element and its negation. Rather than minimizing the difference between these two, dialectic maximizes that difference and brings it to the fore as the basis of a profound and ironic reconciliation. The positive really loses itself in the negative: subject loses itself in object, individual sacrifices oneself for the community, nature contradicts itself by way of freedom. Only by facing its utter self-negation can the positive element come to realize itself in the truest, most definitive sense, which ultimately incorporates the negation into its proper identity. Objectivity becomes integral to the definition of the subject and vice versa. Good and evil turn out to be mutually fundamental and necessary. Reality needs appearance just as appearance needs reality. Dialectic thus overcomes the opposition between positive and negative by making each an intrinsic, defining element of the other. This does not simply nullify or explain away the opposition, for without such opposition there can be no dialectical unity.

Take for example the dialectic between freedom and nature. Hegel begins on the premise that these two really are distinct, different, and even incompatible. Freedom requires the rejection or negation of what is natural (and therefore unfree): "Man . . . must realize his potential through his own efforts. . . . [I]n short, he must throw off all that is natural to him."[15] Freedom thus entails one's refusal to be determined by genes, culture, parents, society, and other external factors and choosing instead to be one's own self. However, paradoxically, because freedom must realize itself by throwing off nature, freedom actually *needs* nature and nature needs freedom. Freedom proves itself to be the very fulfillment of nature; freedom must realize itself by rejecting its negation that is nature, and nature must realize itself by losing itself in freedom.

Dialectic is thus deeply ironic. This makes it difficult to compress dialectic into a short and tidy explanation. The above example is only a rough sketch,

---

15 Georg Wilhelm Friedrich Hegel, *Lectures on the Philosophy of World History: Introduction, Reason in History*, trans. H.B. Nisbet (Cambridge: Cambridge University Press, 1975), 50–51.

and Hegel provides few examples straightforward enough to familiarize a new scholar with the odds and ends of this approach. Surprisingly, Hegel's treatment of the Trinity may represent one of his more accessible examples. Even God is not above this all-encompassing logic. This is because for Hegel, dialectic is not a mere conceptual tool, as though it begins with the philosopher's conscious evaluation of reality. Rather, dialectic is the inner unfolding logic of reality itself—including the very reality of God. To think dialectically is to uncover an inner dynamic that has always already been there.[16]

Rather than being an exception, God serves as the chief embodiment of the dialectical process. When speaking of God as Father, Son, and Holy Spirit, Hegel does not intend this to be understood in any particularly traditional dogmatic manner. He is not speaking of three distinct and coequal "persons" who each and together are the one God. Rather, he reinterprets or—as Cyril O'Regan describes it—"misremembers" this doctrine in order to say something else altogether.[17] For Hegel, Father, Son, and Spirit symbolize three distinct but interrelated aspects of divinity as manifesting itself relative to the world. This Trinity is dialectically structured: Father and Son are two opposite dimensions of divinity, which find themselves reconciled in the Spirit. To understand this, let us look at each of these moments more closely.

(1) God the Father symbolizes the raw Idea of God as universal and transcendent but also thereby enclosed within God's own self. Hegel refers to this first, positive stage as God's *in-itself*. Such a God is only God in principle, not in fact, for this Idea of God is all too abstract and lacking in real content. For example, this God is "love" only in a general sense, because there is not yet an *Other* for God to love.

(2) This too abstract God must be lost within the concrete and particular; God as Son must dive into the reality of the world and become God *for-itself*. In the first place, this movement of God into the world is actually the very creation of the world; God creates the world as other-than-God in order that God might have that other-than-God as a space in which to manifest God's self.[18] In revealing God's self into the world as Son, God succumbs to the negativity or otherness of the concrete, particular world (the Son "becomes

---

16 Hegel, *Phenomenology of Spirit*, §122.

17 Cyril O'Regan, *The Anatomy of Misremembering: Von Balthasar's Response to Philosophical Modernity* (Chestnut Ridge, NY: Crossroad Pub., 2014).

18 Hegel, *Hegel's Philosophy of Mind*, 177.

wickedness" [2 Cor 5:21]). By that same motion, however, God as Son overcomes this self-alienation by "putting himself in judgment and expiring in the pain of *negativity*."[19] Jesus's death on the cross thus becomes not a historical event but rather a symbolic representation of God's own descent into particularity and individuality. By negating God's abstract universality, God as Son embraces concrete history as the sphere of God's own coming-to-be *as* God.

(3) The sum result is that the history of this world turns out to be integral to God's own immanent process of coming-to-be. God becomes more truly God by losing God's self within the world and rediscovering that self as Spirit in the fullest sense. The negative—the other-than-God, which is concrete, historical reality—turns out to be an integral factor in the self-realization of God as Spirit. The God as Son who "suffers" his self-manifestation in the world turns out to be "just as immediate an expression of him as he is himself; he knows himself and contemplates himself in it—and it is this self-knowledge and self-contemplation which constitutes the third element, the Spirit as such."[20] The *in-itself* reaches out *for-itself* and in this way becomes *in-and-for-itself*. God is realized as Spirit, which is universal and concrete at the same time. The universality of Spirit is no longer rooted in abstraction or generality; God as Spirit is irreducible both to a mere abstract notion of divinity and to any one isolated reality within the world.[21] Father and Son are narrow, limited ways of looking at God. Only Spirit is God in the most definitive and fully realized sense.

Two distinct movements intertwine these three dialectical stages: first, the movement from positive to negative (Father to Son); second, the movement from opposition to reconciliation (to Spirit). Hegel refers to this latter, reconciliatory movement with the German word *Aufhebung*. The noun *Aufhebung* derives from the verb *aufheben*. Though one might literalistically translate this verb as "to heave up," its actual meaning varies dramatically according to context. English displays a similar fluidity of meaning in the phrase "hold on!". Only context can tell whether "hold on!" means "stop" or "keep going." In its most basic usage, *aufheben* means "to raise up," e.g., off the floor. By extension, however, it can also mean "to maintain or preserve," just as an appeals judge might "uphold" an earlier ruling. But in another

---

19 Hegel, *Hegel's Philosophy of Mind*, 178.
20 Hegel, *Lectures on the Philosophy of World History*, 51.
21 Hegel, *Phenomenology of Spirit*, §19.

context, *aufheben* can mean the very opposite, "to nullify or abolish," just as in English one might "raise" a blockade or "lift" a restriction.

"To raise up," "to maintain or preserve," "to nullify or abolish"—*aufheben* means all of these things, and Hegel intends them all. The conceptual overlap among these three denotations creates a fluidity of meaning that allows Hegel to express the ironic logic of dialectic. *Aufhebung* is both preservation and nullification. It *nullifies* the very opposition between positive and negative—between God *in-itself* and *for-itself*—but in doing so it actually also reconfirms and *maintains* this same opposition, taking it in as the very medium by which the positive comes to transcend its own limitations. Its preservation is nullification and its nullification is preservation. By analogy, consider how the book of Genesis says that man and woman become "one flesh" (Gen 2:24). The only way that they can become one is because they are initially two. In a similar way, dialectic means that the reconciliation occurs on the basis of the irreducible difference between positive and negative; it both nullifies this difference and maintains it at the same time. In order to highlight this multiplicity of meaning, translators often render *Aufhebung* using a technical term derived from Latin: *sublation*.

Aufhebung is thus the dialectical movement of reconciliation, wherein the negative is revealed to be the basis of the positive's real and truest meaning. It is that climactic event wherein the world turns out to be just as fundamental to God as God is to the world. Nevertheless, to be clear, although *Aufhebung* can be seen as the second movement—from opposition to reconciliation—Hegel does not mean to imply that it occurs discretely *after* the first, original movement from positive to negative. Rather, this second movement properly unfolds as part of the first; it is the inner truth that arises out of this self-negation as its own definitive accomplishment and inversion at the same time. Hence, in the example of the Trinity, the Spirit arises out of the motion from Father to Son, not as a secondary or subsequent turn of events.

The Process of Spirit by Way of History

Because Spirit represents God's most completed aspect, Hegel is more interested in God as Spirit than as Father or Son. In fact, "Spirit" is a weighty word, which signifies more than any one partial aspect of God. After all, the Bible says, "God is Spirit" (John 4:24). Spirit represents the fullness of divinity as such. Yet in order to realize this fullness, God must face God's own opposite; the infinite and eternal ideal must lose itself in finite and historical reality. In this way, Hegel dramatically rehashes the traditional Christian concept of spiritual existence, making it essentially dialectical. It becomes defined according to the conflict of opposites.

The patristic understanding, building upon first-century Judaism and imports from Platonic and Stoic philosophy, saw spirit as something essentially beyond physical reality. God had no body, and thus God's self-manifestation in the physical world was entirely gratuitous and miraculous. As Christians maintained that the material, bodily world was a good creation of this spiritual God, the ontological difference between God and the world served not merely to safeguard God's transcendence but all the more to radicalize this sense of divine gratuity. God did not have to make the world. Therefore, the world exists because of God's good pleasure. It was neither the byproduct of an unavoidable disorder within God's own self (as the Gnostics taught) nor the outcome of God's eternal and unwinnable struggle with some primordial negative reality (as the Manichees taught). At the same time, the difference between God and the world was in some sense created precisely so that God might bridge the gap. For this purpose, humans were created to occupy a privileged place between God and the material world. As finite spirits, we mediated between the material and the immaterial. In fact, for Origen, the natural and proper unity between body and soul served to enable the Son's incarnation. Yet while finite spirit—the human soul—was in some sense defined according to its relation to infinite spirit—God—the reverse was not true. God as spirit was not defined by the difference between God and the world. Prior to the world's existence, God was already spirit in the fullest sense. Spirit preceded and in every way transcended that which was not spirit; the infinite was prior to the finite; eternity existed before the first inkling of time.

By making the world into the site of God's own self-negation, Hegel effectively collapses the real difference between God and the world. God and the world become mutually defining. In order to become God in the truest sense, God must reveal God's own self into the world. For example, God cannot merely be Love in principle or interiorly but must actually love the world and thus reveal God's self as Love by relating to the world. In this way, Hegel's God is *necessarily* self-revealing. He writes, "That God is spirit consists in this: that God is not only the being that maintains itself in thought but also the being that appears, the being that endows itself with revelation and objectivity."[22] Not only must God create the world in order to become God, but God's self-revelation within the world becomes an unavoidable and innate function of God's character as Spirit. Just as the body is the self-expression of the human soul, so also the world is the external manifestation of divine Spirit.

The centrality of revelation in Hegel's philosophy hints at one of Hegel's key sources: Neoplatonism. Ludwig Feuerbach famously called Hegel "the German Proclus."[23] This moniker only gestures at the monumental significance of Neoplatonism within Hegel's thought. Plato believed that the physical world existed in order to manifest the beauty and diversity of the world of Ideas. The Neoplatonists such as Plotinus and Proclus took up this claim and radicalized it. They saw the self-revelation of the Ideas as integral to the Ideas themselves; that which is true and eminently real *must* reveal itself in concrete, external existence. In effect, the divine Ideas must prove their preeminent beauty and rationality by manifesting themselves in the beauty and rationality of the world. For Hegel, this will ultimately mean that God must become truly God by manifesting God's self within the world. Self-revelation is a divine act of coming-to-be, and the world is the necessary sphere of God's concrete self-realization.

For Hegel, the world externalizes or reveals God not only in its materiality but also in its spiritual dimension—the Platonic "world soul." Hence God manifests God's self in the material world precisely because that materiality is the outward, visible dimension of the Spirit of the world. The Spirit of the world, epitomized by human consciousness, is the other side of the divine Spirit; it is God enmeshed within physical reality and in the process of realizing God's self through this dialectical struggle. As humanity comes to maximize its spiritual consciousness in the world, this same process is that whereby God becomes more truly Spirit. After all, for Hegel the divine Spirit is by definition the very self-consciousness of God, and this self-consciousness is actualized by humanity's ever-advancing knowledge of God: "God is God only so far as he knows himself: his self-knowledge is, further, his self-consciousness in man, and man's knowledge *of* God, which proceeds to man's self-knowledge in God."[24]

---

22 Georg Wilhelm Friedrich Hegel, *G. W. F. Hegel: Theologian of the Spirit*, ed. Peter C. Hodgson (Minneapolis, MN: Fortress Press, 1997), 176.

23 Ludwig Feuerbach, *Principles of the Philosophy of the Future*, trans. Manfred H. Vogel (Indianapolis, IN: Bobbs-Merrill, 1966), §29.

24 Hegel, *Hegel's Philosophy of Mind*, 176.

The dynamism of this God–world relationship sets Hegel's view apart from any run-of-the-mill pantheism. Hegel explicitly disavows such pantheism as found in Hinduism or the philosophy of Baruch Spinoza, because these treat the unity of God and the world as something merely given beforehand and apart from any dialectical process. God and the world are from the beginning one and the same thing looked at from difference perspectives. Diversity or difference thus appears to be a mere veil over this primordial unity, so that no matter how distinct we may seem to be, we are all really integral parts of one greater original whole. Such bland pantheism not only lacks any real growth or progress, it also fails to account for the diversity or disunity of the world in any meaningful way. Gandhi and Stalin turn out to be two incidental pieces of one and the same amorphous whole.

In contrast, Hegel's pantheism is dynamic. The identity between God and the world intensifies over time as it moves from the mere implicit unity of existence, to the diversity of concrete experience, unto the comprehensive dialectical unity that incorporates unity and difference at the same time. Initially, the world is Spirit only in a lesser or potential sense; Spirit is the world's unfulfilled ideal. The world comes to be self-identical with Spirit in the fullest sense only by means of the world's historical process.

Process as Creativity

From this brief overview, we can see that Hegel's philosophy proposes a comprehensive and systematic understanding of God as making progress through the ongoing history of the world. God's very existence follows the logic of dialectic, which thus makes the conflict of opposites integral to God's very identity. God is as dependent upon the world as the world is upon God. Ultimately, the destiny of the world is to become Spirit—God—in the fullest sense through the perfection of human consciousness. Such a perspective dramatically rethinks the relationship between God and the world precisely by reconfiguring the metaphysics of being (ontology) and the meaning of creativity.

Creativity, as we saw in the previous chapter, would become a central concern for nascent evolutionary theology around the turn of the twentieth century. Henri Bergson's philosophy exemplified how many philosophers and theologians sought to reintroduce an element of creativity into the dysteleological process of evolution. For his part, Bergson saw the world as driven by an inner vital impulse (*élan vital*), which allowed it both to develop naturally and at the same time produce something new—something not simply reducible to a prior and implicit condition. In contrast, a purely mechanical, noncreative world might be created by God at the beginning and wound up like a watch. Anything occurring after this beginning would be no more than the logical and necessary consequence of that initial creation. Such a world would be like a log rolled down a hill. It might move this way and that, but its movement would be entirely determined by the shape of the log and the slope of the hill. Seeking to avoid such mechanical determinism, Bergson saw the world as created in such a way as to include some element of unpredictability within it. The vital impulse drives evolution's movement in a way that exceeds a mere mechanical unfolding. The world proceeds in a meaningful direction, but the particulars of its movement are not set in stone. By analogy, such a world is like a soapbox racer with a driver; its movement is indeed determined by the slope, but the driver's influence and shifting ballast can dramatically impact how its movement unfolds.

Hegel's dialectical logic serves much the same purpose, but it does so precisely by locating God's *own* creativity within the sphere of the world. Bergson certainly was not interested in limiting the world's creativity, but he still saw the world's evolution as something distinct and altogether separate from God's own creative action. In fact, he wanted to stress that the vital impulse was not a constitutively external influence such as God's ongoing direct intervention. This would still make the world mechanically bound to follow some predetermined path, even if this influence extended beyond the first moment of creation. Hegel overcomes such a problem, however, precisely because he dialectically annuls the very difference between God and the world. The unpredictable element in the world's evolutionary process is not a mere shadow, similitude, or second relative to God's free creativity. Rather, it is the very creativity of God in its finite process of becoming. Creativity is the manifestation of God's freedom over and against predetermined, necessary, and mechanical nature. Spirit, which makes progress through the world and its history, is the real unfolding of such freedom and is therefore creativity itself.

Here we see the pivotal link between Hegel and Whitehead. In fact, their relationship goes far deeper than a few theological commitments; they rely on the same fundamental dialectical logic. As George Lucas Jr. puts it, for both Hegel and Whitehead "God is a necessary and integral feature of the more general cosmology which God enables, and in which God participates (or, more correctly, through which God is more fully actualized)."[25] The creation of the world is an essential element in God's own coming-to-be, such that history and theology are effectively two sides of one and the same dynamic process. C. Robert Mesle summarizes Whitehead's view in this way: "Creativity is not a mysterious power *behind* the world that makes it go. It is not fate, destiny, or a power that works for either good or evil. Creativity is simply the ultimate feature shared by all that is actual—God and the world alike."[26] In this sense, creativity is really prior to and more fundamental than God.[27] Creativity requires God only in the sense that God is its prototypical embodiment.[28] God is like creativity's original proof of concept. Nevertheless, creativity is not itself some higher God but rather the inner logic by which existing realities strive toward higher excellence.

In fact, notwithstanding some differences in terminology, emphasis, and approach, what Hegel refers to as "dialectic" is effectively the same logic, function, or process that Whitehead names *creativity*. Just as Hegel sees dialectic as the reconciliation of opposites through the annulment (sublation) of difference, Whitehead sees creativity as the complex and cyclical process, energized by the appearance of "contrasts," by which existence advances from "disjunctive disunity" to conjunctive unity or "the production of novel togetherness"—a production that Whitehead terms *concrescence*.[29] Creativity is "the ultimate metaphysical principle." It a necessary, primordial, and intrinsic element of all existence, even the very existence of God. Like Hegel's dialectic, Whitehead's creativity is more fundamental than Kant's categories, more universal than the scholastic transcendentals, and more intimate than the Aristotelian form or essence. More importantly, by hinging development

---

25 Lucas Jr., *Two Views of Freedom*, 3.

26 C. Robert Mesle, *Process-Relational Philosophy: An Introduction to Alfred North Whitehead*, electronic resource (West Conshohocken, PA: Templeton Foundation, 2008), 80.

27 See Alfred North Whitehead, *Religion in the Making: Lowell Lectures, 1926* (Cambridge: Cambridge University Press, 1927), 141.

28 Alfred North Whitehead, *Process and Reality: An Essay in Cosmology*, ed. David Ray Griffin and Donald W. Sherburne, Corrected ed. (New York: The Free Press, 1978), 7 (10–11). The first set of numbers refers to the pages of the original 1929 Macmillan edition, which this newer, corrected edition provides in square brackets. The second set of numbers, in parentheses, refers to the corrected edition itself.

29 Whitehead, *Process and Reality*, 32 (21).

upon the reconciliation of "contrasts" or opposites, Hegel and Whitehead both make duality or difference fundamental to the world and its definitive unity. The process of coming-to-fullness, whether we name it dialectic or creativity, requires such dualities as permanence and change, mind and body, order and chaos, God and the world, and even good and evil, precisely so that these may become dynamically and productively united.[30] Diversity stands at the base of ultimate unity.[31] A mere pre-given unity has never faced its opposite, which challenges its very legitimacy. Such a bland unity thus produces nothing but boredom and stagnation. True novelty or progress comes about only by facing contrast, by walking through the fire, by meeting ultimate negation and facing the "pain" of chaotic dissolution.

Theologically, the sum result for Whitehead as well as Hegel is that God, in order to become truly God, must face utter dissolution and loss of self by intimately descending into the concrete, objective world—God's very own opposition or negation—in order thus to rediscover God's self as the definitive unity of the world and its God, of matter and meaning, or of the concrete and the universal. God needs the world in order to even become God in the truest sense. For Whitehead, even as God serves as the "instrument of novelty" for the world, the world serves as the same for God. In this way, the logic of creativity or dialectic is more fundamental, more primordial than God's own self; there could be no God without it.

WHITEHEAD'S PROCESS PHILOSOPHY

Having explored Hegel's dialectic, we are now well-positioned to recognize the broader significance of Whitehead's point of view. The resemblance between these two systems is more than coincidental. Nevertheless, this does not make their relationship a simple story of direct inheritance. Whitehead purportedly did not read Hegel to any significant extent when formulating his own ideas. The formal differences between them—differences for example in terminology and method—highlight some of the ways in which intervening influences and conditions contribute to the unique flavor of Whitehead's process thought.

---

30 Alfred North Whitehead, *The Function of Reason* (Princeton, NJ: Princeton University Press, 1929), 25.
31 Whitehead, *Process and Reality*, 528–30 (348–49).

Rather than imitating Hegel himself, Whitehead derived his Hegelianism through the influence of the British school of absolute idealism, which rose to prominence in the latter half of the nineteenth century. Spearheaded by F. H. Bradley (1846–1924) and Bernard Bosanquet (1848–1923), this school of thought shaped the philosophy of Samuel Alexander (1859–1938), a contemporary and significant influence for Whitehead.[32] This intellectual lineage did more than simply transmit Hegel to Whitehead. It also set much of the tone of Whitehead's philosophical project as a dialogue with and critique of British empiricism.

From its inception, British idealism was always an amalgam. It developed within an intellectual milieu marked by the ongoing influence of the empiricism of John Locke (1632–1704) and David Hume (1711–76). British philosophers adopted Hegelian ideas in order to counteract what they saw as empiricism's accelerating decline into materialism. Yet though critical of this decline, they did not reject empiricism wholesale. Rather, their questions, key concerns, and methods of approach owed much to the empiricist tradition. In this way, they aimed to achieve a higher synthesis of empiricism and idealism. They sought a philosophical standpoint that would incorporate each tradition into a definitive, all-encompassing, and systematic view of reality.

Whitehead's own attempt at such a synthesis leans heavily on characterizing the ongoing process of the world in terms of creativity in order thereby to tie this evolutionary development to a sense of deeper meaning. In what follows, a brief analysis of Whitehead's philosophical quest for meaning will enable us to better understand his most central metaphysical concepts. These form a conceptual framework that decisively reconfigures traditional Christian understandings not only of the world but also—quite intentionally—of God's own self. Integrating God into a progressive (and in that sense evolutionary) framework of meaning ultimately requires that God, too, become dependent upon the workings of this process in order to obtain God's own meaning and fulfillment. As in Hegel, Whitehead's dialectical logic (creativity) comes to be more fundamental, more absolute than God. Although Whitehead originally dubbed his system "the philosophy of organism," it is widely known today as *process theology*.

---

32 See Alfred North Whitehead, *Science and the Modern World* (Cambridge University Press, 1925), xi; *Religion*, 100.

Metaphysics and the Recovery of Meaning

Whitehead wrote in an intellectual milieu in which British empiricism seemed to have drained the world of real, transcendent meaning. Interestingly, his response to this problem arose far afield from the domain of speculative philosophy. Born in England in 1861, he spent his early career lecturing primarily on mathematics. Inspired by the abstract, systematic, but also heuristic character of this discipline, Whitehead would later argue that modern philosophy and science owe their origin to the rise of modern mathematics and its interest in understanding abstract, general principles.[33] Without reducing philosophy to mathematics, Whitehead would see philosophy's real significance in its ability to move beyond empirical occurrences toward an understanding of the general principles, which illuminate the meaning of such occurrences.[34] It was not until his late fifties, however, that Whitehead undertook philosophy in a professional capacity. Some early works on the philosophy of science resulted in an offer to join the philosophy faculty at Harvard, which he accepted in 1924. Whitehead's philosophy flourished in this new position. He expanded into the realm of metaphysics, developing his process-based framework of reality and cognition.

The critique of modern philosophy and its tendency toward meaningless materialism serves to some extent as a unifying thread within Whitehead's thought. As he sees it, the laudable success of modern science and mathematics via influences such as Newton's physics and Hume's empiricism have borne an unfortunate desiccatory effect on modern thought. Its emaciated view sees reality as a collection of material bodies and forces without any overarching rhyme or vivifying reason. In science, we are left with a broad array of facts that can be categorized but not really understood. The physical world remains essentially unintelligible. To use a present-day example, science can extrapolate back to the existence of a Big Bang, but it cannot provide any reason for why such an event should have occurred or what it means for the world as such. Newton "thus illustrated a great philosophic truth," writes Whitehead, "that a dead Nature can give no reasons. All ultimate reasons are in terms of aim at value. A dead Nature aims at nothing."[35] Kant enshrines this toxic

---

33 Whitehead, *Science and the Modern World*, 29–56.
34 Whitehead, *Process and Reality*, 12 (8).
35 Alfred North Whitehead, *Nature and Life* (Cambridge: Cambridge University Press, 1934), 23–24.

inheritance as a fundament of philosophy by restricting our knowledge to the realm of mere appearances or *phenomena*.[36] Kant's "pure reason" remains largely impotent to know whether such phenomena have any deeper basis or meaning.

In response, Whitehead argues that the real, productive genius behind science and philosophy has always stemmed from Pythagoras and Plato.[37] As mathematicians, they were able to recognize how concrete, empirical experience reveals broader logical and harmonious interrelations. In this way, Plato's mathematical framework led directly into his metaphysics. To paraphrase Hegel, idealism is the conviction that reason is coextensive with all reality.[38] The reality that we experience, the world of phenomena, is therefore shaped through and through by an inner rationality; it makes sense. For Plato, the self-subsistent and eternal Ideas manifested their beauty and rationality in and through concrete, material realities. The beauty of "Motherhood" was made manifest through the reality of actually existing mothers. One could say, therefore, that reason's self-embodiment within concrete realities was productive of meaning or value. In fact, this necessary but dynamic interconnection between reason and empirical reality becomes all the more pronounced in Neoplatonism and, through it, the philosophy of Hegel.

Consequently, Whitehead sees a revival of metaphysics as the necessary cure for modernity's loss of meaning. Just as "you cannot shelter theology from science, or science from theology," neither "can you shelter either of them from metaphysics, or metaphysics from either of them."[39] Metaphysics goes hand in hand with the recognition that real existence implies the attainment of value.[40] Things exist in reality so that they may achieve higher value through the process of creativity. Hence, inasmuch as Whitehead sees evolution as a teleological and purposive process—he rejects any rigorous, dysteleological Darwinism—he also understands evolution as the defining characteristic of existence as such. Real existence *is* the evolutionary process of creativity.

Our next step is to understand how Whitehead's metaphysical language not only encapsulates this perspective but also implicitly leads to a rethinking of the nature of God. By emphasizing the unity of reason and its orientation

---

36 Kant, *Critique of Pure Reason*, xxxv.
37 Whitehead, *Science and the Modern World*, 40–43.
38 See Hegel, *Phenomenology of Spirit*, 140 (§233).
39 Whitehead, *Religion*, 67.
40 Whitehead, *Function*, 24.

toward the question of meaning, Whitehead not only implicitly follows the pattern of thinkers like Teilhard de Chardin and Bergson, but he also makes the link between process philosophy and theology inevitable. The question of the meaning of the world as such really *is* the question of God in its inverse form. Thus while many specific points of theology will be developed by Whitehead's followers, Whitehead himself will provide the key insight: History is the process of God's self-realization in and through the ongoing act of creation.

Key Metaphysical Terms

Whitehead's innovative and often alluring metaphysical terminology contributes both to the appeal of his system and to its incumbent difficulty for new readers. Like Hegel, Whitehead frequently forms his own philosophical jargon by infusing common words with new meaning. Whitehead shows particular fondness for emotive, existential, and psychological terms such as "feeling," "lure," "urge," and "enjoyment." This preference stems not only from the influence of philosophers such as F. H. Bradley and William James (1842–1910) but also from that of English Romantic literature.[41] Hegel sometimes uses comparatively evocative terminology too.[42] However, such language takes center stage in Whitehead's metaphysics. His *logos* or logical system is suffused with a strong sense of *pathos*—a kind of emotive energy, one might say, which makes it irreducible to a cold, mechanical logic.

Whitehead's emotive terminology serves to highlight how creativity operates even prior to explicit consciousness. For example, a word such as "feeling" is already useful at its surface because it does not necessarily imply rational, cognitive experience. One can feel offended whether or not reason dictates that this should be the case. Feeling is not *logocentric*—to use a term popular within contemporary postmodern philosophy. It is not reductively rational. Whitehead uses this terminological shift away from rationality to push back against the emphasis on consciousness that has dominated philosophy since the time of René Descartes. In contrast, for Whitehead, *feeling* does not imply

---

41 Whitehead, *Process and Reality*, xiii (vii–viii). See Antoon Braeckman, "Whitehead and German Idealism: A Poetic Heritage," *Process Studies* 14, no. 4 (1985): 265–86.

42 See for example Hegel's discussion of consciousness as "the enjoyment of pleasure." *Phenomenology of Spirit*, 218.

a conscious experience. Rather, feeling defines a being or entity as such, so that even a rock or a sticker has feeling in the Whiteheadian sense. To be clear, Whitehead does not pretend that stones are subjectively conscious. As Ian Barbour explains, by attributing "at least *rudimentary forms of experience*" to unconscious existents, Whitehead challenges us "to look at the world from the viewpoint of the entity itself, imagining it as an experiencing subject."[43]

Whitehead draws this surprising usage of "feeling" from Francis Bacon (1561–1626), who speaks of all existing things as having "perception."[44] This includes unconscious or insensitive existents such as sticks, mud, and atoms. Bacon's "perception" serves to describe the way in which existents passively receive the influence of one another. Iron "perceives" the influence of a magnet is thus drawn toward it. Such inanimate perception even transcends the limits of conscious sense-perception: "In truth, the air so keenly *perceives* hot and cold that its *Perception* is far more subtle than human touch, which is nevertheless taken to be the measure of hot and cold."[45] In much the same way, Whitehead believes that metaphysics reveals being's orientation toward meaning as operative even prior to the advent of rational consciousness. Inasmuch as all existents are directed toward dialectically realizing the Ideal through the process of creativity, even unconscious existents have a "blind urge towards a form of experience, that is to say, an urge towards a form for realization."[46] Even the primordial production of the universe's subatomic particles was driven by a kind of inner, unconscious urge toward development.[47]

Existents are not drawn toward development merely as isolated individuals but rather as part of the complex and interrelated matrix of existence. Whitehead substitutes a new word—*prehension*—for Bacon's "perception."[48] "Prehension" refers specifically to the physical interrelation of existing things, which for Whitehead is so intrinsic and indispensable to existence that particular existents cannot really be thought of apart from the experience of

---

43 Ian G. Barbour, *Religion in an Age of Science*, 1st ed. (San Francisco: Harper and Row, 1990), 223, 226. See Whitehead, *Process and Reality*, 130–31.

44 Francis Bacon, *Sylva Sylvarum: Or, a Naturall History in Ten Centuries*, 6th ed. (London: William Lee, 1651), 171.

45 Francis Bacon, *De augmentis scientiarum libri IX* (Amsterdam: Henrik Wetstein, 1694), 256–58 (book 4, chap. 3). My translation.

46 Whitehead, *Function*, 26.

47 Whitehead, *Function*, 19.

48 Whitehead, *Science and the Modern World*, 100–101; Conrad Bonifazi, *The Soul of the World: An Account of the Inwardness of Things* (Washington, DC: University Press of America, 1978), 202. Cf. *Process and Reality*, 81–83 (52–53).

prehension. To modify Bacon's example: the iron *prehends* the magnet; its very essence as iron is qualified by its magnetic potentiality. For Whitehead, prehension is not merely one aspect of being among others. Rather, reciprocal and ubiquitous prehension characterizes all concrete beings as such. The magnet prehends the iron even as the iron prehends the magnet. Both prehend the broader universe of which they are a part. Put another way, within the world "each part is something from the standpoint of every other part, and also from the same standpoint every other part is something in relation to it."[49] Each existent only makes sense in relation to its environment—that is, to its being contextualized by every other existent to which it relates by way of prehension. To think of a magnet apart from its relation to iron, for example, would be a mere abstraction; it could have no truth.

In this way, Whitehead pushes for an ontology—a philosophical understanding of the nature of being—that differs fundamentally from the mainstream. Commentators often present Whitehead's ontology as a rejection of Aristotle. In truth, though it cannot be treated here in full, Aristotle's ontology is not as static or anti-relational as this rejection would suggest. Whitehead is less concerned with Aristotle himself than with how his ontology has been refracted through modern thought—and through Descartes in particular.[50] Still, even if Aristotle does not stand in strict opposition to Whitehead, there is a clear and real difference in emphasis between the two. Accordingly, even a rough and imprecise outline of Aristotle's ontology will serve to illuminate Whitehead's position in relief.

For Aristotle, the basic unit of reality is *substance* (*ousia* in Greek). This term has many different meanings throughout his corpus.[51] For our purposes, let it suffice to say that an individual existing being is a substance. When taken in this sense, for example, a cat is a substance. It is a unity of form and matter, or, to put it another way, it is a realized instance of the form of catness: whiskers, purring, the whole package. What is truly *substantial* and therefore indispensable to the particular existent is that which remains fundamental and unchanging. A cat can be shaved without ceasing to be a cat; therefore, hair is not strictly speaking a substantial quality to a cat. In this way, that which is accidental—a quality that may or may not pertain to a cat—is inherently less real. An accident such as orangeness or fluffiness does not exist on

---

49 Whitehead, *Science and the Modern World*, 95.
50 Whitehead, *Process and Reality*, 81 (51).
51 See Aristotle, *Categories* §1.1.5; *Metaphysics* Γ.2, Δ.8, Z.1–3, 11.

its own but must borrow the existence of the substance. That accident may change or cease to exist. The cat may cease to be orange. Yet this has little to no bearing on the existence of the substance itself. An individual cat remains a concrete, particular granule of being despite any accidental alterations. Likewise, Aristotelian ontology tends to consider the relationship between various substances more as something accidental than substantial. Only because a cat first and foremost exists can it then be considered to exist in relationship with its mother, its owner, or its favorite couch.

In contrast, for Whitehead, to think of an existent in terms of substances is to picture a mere abstraction. Such a conceptualization implies that the meaning of a cat remains the same even without reference to its role in the broader ecology of the world. In order therefore to reconfigure the way that metaphysics conceptualizes being as such, Whitehead coins a new way of speaking about its individual constituents.

For Whitehead, an individual existent or granule of being is an *actual entity* or *actual occasion*.[52] Despite how "actual" typically functions in English, such an entity is not "actual" as opposed to "simulated" or "fake." The philosophical meaning of "actual" is better illustrated by its use in French, Spanish, and German to indicate "currently active" or "up-to-date" as opposed to "past" or "outdated." Similarly, in medieval, scholastic philosophy, a thing is "in act" when it exists in the present in a real and tangible manner as opposed to being a mere unrealized possibility.

Whitehead sees the actuality or present existence of an individual existent as constituted through and through by its relationships with others. Relationality is the very stuff of being. For example, even a cat's most distinctive characteristics, such as its predatory nature or its ability to land on its feet, derive from its relationships to others, including its feline forebears, its environment, its predators, and its prey. Analyzing an actual occasion in terms of its prehensions illuminates how its very being is constituted by its many and diverse relationships. Note that "analyze" derives from the Greek *analuō*, "to thoroughly untie." Prehensions are like the individual fibers spun together tightly into an intricate length of yarn. Analysis mentally separates these fibers in order to perceive how the actual occasion is constituted by its many prehensions.

---

52 Whitehead, *Process and Reality*, xiii, 18 (viii–ix, 27–28). These two terms are almost always interchangeable except in regard to God (see below). Note that in *Religion in the Making*, Whitehead instead uses "actual entity" and "epochal occasion." *Religion*, 78.

By prioritizing relationality, such an ontology implicitly understands being as fundamentally open to influence and change. Traditional ontology often thinks of being in terms of *identity* or selfsameness (from the Latin *idem*, "the same"). Roughly speaking, the being of a cat remains the same whether that cat is wet or dry, here or there, hungry or full. This focus on identity deemphasizes relationship because, for example, the being of the cat remains the same whether it is my cat or your cat. In contrast, an actual occasion is fundamentally defined by its relationships. Its very identity—its selfsameness—is a dynamic process of *prehensive unification*. It is the coming together of various elements by way of prehension. Every prehension is a relationship, a mutual situation of determining and being determined by a concrete other.

Inasmuch as prehensive relationships are fluid and dynamic, an actual occasion is by its very nature continuously in process. This process is driven by the aforementioned urge toward self-formation. This is not a blind inclination toward achieving some divinely mandated final cause. The progressive development of an actual occasion is not determined by any external, preordained purpose but rather by its own inner *subjective aim*.[53] One might say that the subjective aim represents the intrinsic, rational self-ordering of the actual occasion toward the production of something new. It is the actual occasion's *feeling* of its own purposeful direction, though "feeling" here again does not imply conscious emotion. A piece of iron has a subjective aim akin to that of a human being.

In this way, an actual occasion is inherently self-creative. Creativity is its origin and its *telos* (goal). *Concrescence* is Whitehead's name for the process by which an actual occasion creatively produces itself out of the union of prehensions.[54] For example, a cat exists because of its relationships—including for example its relation to its parents and environment—yet in Whitehead's view, the cat's act of existence is not thereby merely passive or receptive. Rather, the cat is the subject of its creative coming-to-be. In fact, its very process of becoming defines it as an actual entity. Its concrete reality is not its unchanging catness but rather its fluid process of becoming a cat. At the same time, each actual occasion contributes to the matrix of prehensions, serving as the material for the concrescence of other actual occasions. The cat, by coming-to-be, contributes to the environment or context in which

---

53 Whitehead, *Process and Reality*, 25–26 (37–38).
54 Whitehead, *Religion*, 89; *Process and Reality*, 321–22 (211).

other cats and other beings also come-to-be. Creativity is thus a restless movement—the ceaseless advance of concrescent reality.

Although Whitehead thus defines existence according to the continuity of becoming, his metaphysics is also essentially *atomistic*.[55] It imitates quantum physics, which sees forms of energy as existing at one and the same time as a continuous wave and as a trail of individual particles.[56] The magnitude of energy is both continuous and discrete. Light is both wave and particle. In the same way, an actual occasion is both a discrete existence at a particular point in time and, only in view of this, also an ongoing entity whose fuller reality transcends any particular moment of its existence. In the first sense, an actual occasion is like a slice of reality. It functions in a way similar to an inertial frame of reference in physics. A cat *at this particular moment* is an actual occasion and therefore a unique concrescence of prehensions, which is uniquely related to *this* world as a whole. In this sense, even the slightest change would alter its relation to the world and make it something other than what it was. In the second sense, however, an actual occasion is also what Whitehead terms a *superject*; it also stands above every particular moment of existence and constitutes something continuous.[57] By analogy, in the first (atomistic) sense, we might talk about one cat existing at 1:00 p.m. and another, different cat existing in the same spot only a moment later. In the second sense, it is one and the same cat who, as a unity of discrete moments of becoming, is always an ongoing process of becoming; this superject is irreducible to any single moment in this process.

This brief exploration of Whitehead's key terms illustrates at the very least the inherent complexity involved in understanding his system. Since Whitehead—like Hegel—dialectically integrates opposites, even his most categorical definitions may at times be viewed in a completely contradictory manner. Anything short of a complete examination can easily produce misconceptions. As shown above, one might easily recognize the continuity of an actual occasion while overlooking its discrete, atomistic character. Unfortunately, a full examination of Whitehead's metaphysics would far exceed what can reasonably be undertaken within this brief overview. Nevertheless, the ultimate unifying element in Whitehead's thought remains the dialectical logic he terms creativity. By recognizing that for Whitehead all truth consists

---

55 Whitehead, *Process and Reality*, 28–29, 53–54 (19, 35–36).
56 Whitehead, *Process and Reality*, 365 (238–39).
57 Whitehead, *Process and Reality*, 43, 71–72 (29, 45).

of the integration of opposites or contrasts, one can surmise—and correctly so—a great many details that could not be treated here, including for example that there are both positive and negative prehensions;[58] that becoming is an interplay of freedom (creativity) and necessity ("givenness");[59] and that actuality integrates both a physical pole and a mental pole, which roughly correspond with the past/real and the future/Ideal.[60] This same logical interplay of dualities will lead directly into Whitehead's theology.

Creativity and God

As we have seen, Whitehead's metaphysics reconfigures our understanding of being in terms of process and relationship. In place of substance, Whitehead identifies the actual occasion (or actual entity) as basic unit of existence. Actual occasions are analyzable in terms of their prehensions, which are the interrelationships with other actual occasions that effectively constitute the actual occasion. The process by which prehensions creatively develop new actual occasions is termed concrescence, and this process is urged forward by the actual occasion's inner subjective aim—its teleological orientation toward the creation of itself.[61]

What remains in this brief overview is to understand how and why this evolutionary metaphysics necessarily reconfigures our understanding of God. God is no peripheral topic for Whitehead. Rather, since he believes (following Hegel) that only a complete and all-encompassing philosophical system can have any truth, he also believes that this ontology must necessarily apply to God in order for it to apply at all. The problem with the substance-based ontology of Descartes, for example, was that by making mind or spirit essentially unrelated to body or matter, it grounded the significance of mind on its being exceptional—i.e., independent of the external world. All the more and in the same way, God served as the ultimate metaphysical exception. The world may be finite, transitory, and dynamic—but *not* God. In Whitehead's view this exceptionality has not served to keep philosophy within its proper

---

58 Whitehead, *Process and Reality*, 35, 66 (23–24, 41–42).

59 Whitehead, *Process and Reality*, 67–72 (42–46).

60 Whitehead, *Process and Reality*, 48–49, 423, 528 (32–33, 277, 348); *Adventures of Ideas* (New York: New American Library, 1955), 274–75.

61 Whitehead, *Process and Reality*, 130–31 (85).

limits. Rather, it has improperly discouraged creative speculation and thus provided for the illicit dominance of mere inductive reasoning without any unifying speculative vision.

In response, Whitehead sees God as an actual entity and therefore as understandable in the same terms as other actual entities within the world.[62] Just as with any other actual entity, to think of God apart from God's relation to the world would be a mere abstraction—a myth that lacks any relevance or reality. For Whitehead, God does not exist apart from God's role in the process of the world.[63] Since creativity constitutes the very being of God, only by way of creating does God come into God's own existence.

This is not to say, however, that God is merely one actual entity among others. In fact, while actual entity and actual occasion are usually interchangeable, God alone is an actual entity but *not* an actual occasion.[64] God is not an exception, but God *is* a sort of exemplar.[65] God specially embodies the process of creativity such that one could even say that God really *is* the process of the coming-to-be of the world.

To better understand God's role in this process, it helps to see concrescence from two different angles. On the surface, we see that real, physical things contribute to other real, physical things. A tree is shaped by the forest, the weather, other trees, and so forth. These are *physical prehensions*. Remember, however, that Whitehead's purpose is to reintegrate meaning into reality. A merely physical process of becoming could just as well be meaningless. In order to make this process inherently meaningful, Whitehead argues that actual occasion is also constituted by its subjective aim, i.e., its inner self-orientation toward purpose or meaning. An actual occasion thus reflects something greater than itself. It integrates not only physical prehensions but *conceptual prehensions* as well. An actual occasion is not only an atom or basic unit of existence but also an atom of meaning or, to put it another way, an atom of reality experiencing itself as meaningful.[66]

These two ways of viewing the formation of an actual occasion— (1) in terms of its physical prehensions and (2) in terms of its conceptual

---

62 Whitehead, *Process and Reality*, 18 (28).
63 Whitehead, *Religion*, 94.
64 Whitehead, *Process and Reality*, 135 (88).
65 W. Norman Pittenger, *Alfred North Whitehead* (London: Lutterworth, 1969), 33–34.
66 Whitehead, *Religion*, 87; *Process and Reality*, 72 (45).

prehensions—correspond to two "formative elements" or "poles" of the dialectical process.[67] The coming-to-be of an actual entity is both (1) the creative integration of real, particular, existing beings (prehensions) into something new; and (2) the meaningful self-embodiment of non-real, universal forms or "eternal objects" (Plato's Ideas) within concrete reality.[68] For the sake of simplicity, think of these as (1) reality developing itself and (2) the Ideal revealing itself. These are the two sides of Whitehead's dialectical process. The first perspective, if held in isolation, would result in mere materialism. The second in isolation would be an overly abstract form of idealism. In contrast, a more complete idealism like that of Hegel or Schelling intimately unites these two perspectives. "Idealism is the soul of philosophy; realism is its body," writes Schelling, "only the two together constitute a living whole."[69] After all, the truest dialectical idealism must reconcile or sublate even the difference between the real and the Ideal.

Yet Whitehead's dialectic admits of a third element, namely, God. (1) The first element (the real) is actual, temporal, and particular; (2) the second (the Ideal) is non-actual, nontemporal, and universal; (3) God, the third element, is actual, nontemporal, and in that way universal. God thus serves as the medium by which the first two elements are dialectically united. On the one hand, God's universality means that God encompasses the realm of Ideas and transcends all of history precisely so that these Ideas may become present at every moment and to every actual occasion within history. Whitehead terms this aspect of God, wherein God serves as the presupposition for all creative process, the *"primordial nature of God."* It corresponds with Hegel's God *"in-itself,"* God as a mere abstract and unreal notion prior to and apart from the world.[70] On the other hand, God's actuality means that God is the exemplary actual entity who, like all other actual entities and in unison with them, exists in and as a process of self-becoming. This is the *consequent nature of God*, the sense in which God creatively incorporates all actual occasions into God's own immediate existence.[71] This roughly corresponds to Hegel's God *in-and-for-itself*.

---

67 Whitehead, *Religion*, 77.
68 Whitehead has Plato's Ideas in mind, but he is not in strict agreement with Plato on what qualifies as an Idea or form. *Process and Reality*, 69–70 (44).
69 Friedrich Wilhelm Joseph von Schelling, *Of Human Freedom*, trans. James Gutmann (Chicago: Open Court, 1936), 30.
70 Whitehead, *Process and Reality*, 50, 70 (34, 44).
71 Whitehead, *Process and Reality*, 525, 531–32 (346, 350).

In conclusion, for Whitehead, the nature of God is bound up with the very idea of process. He writes, "The purpose of God is the attainment of value in the temporal world."[72] Because of this, God needs the world as much as the world needs God. Or as Whitehead puts it, "It is as true to say that God creates the World, as that the world creates God."[73]

Although we cannot here discuss the many critiques of Whitehead's position, recognizing the gravity of Whitehead's assertions is vital for understanding its influence on subsequent thinkers. Whitehead is not merely tweaking traditional Christian theology but rather—like Hegel—overturning it at its core.

Augustine embodies the thrust of traditional theology when he describes God as paradoxically both absolutely immanent to the world and absolutely transcendent beyond the world. God is one and the same time both "more inward than my inmost" (*interior intimo meo*) and "higher than my highest" (*superior summo meo*).[74] The incarnation proves God's transcendence by way of immanence; that is, precisely because God exists independently and beyond the world, God is able to enter into this world in the most intimate and even scandalous manner. The Son of God is in no way compelled to become man. Rather, the Son freely chooses to be born of Mary, to suffer, to die, and to rise, out of the purest love. Since God gains nothing from helping us, this love is gratuitous and free of self-interest.

Like Hegel, Whitehead can describe God using many of the same terms and phrases. Yet in the end, Hegel and Whitehead envision a God who is as much interested in God's own benefit as in ours. The God of process does not create the world out of pure gratuity, but rather precisely because this God must create in order to become God in the truest sense. Proponents of process theology will favor this ontology because it makes God so interdependent with the world that God's relationship with creatures becomes fundamental to who God is. Rather than a special event of divine mercy, the incarnation becomes a natural and necessary outflow of God's own being.[75] As we shall see below, this will contribute especially to their response to the problem of evil.

---

72 Whitehead, *Religion*, 87.
73 Whitehead, *Process and Reality*, 528 (348).
74 Augustine, *Confessions* III, 6, 11.
75 See for example Beatrice Bruteau, *God's Ecstasy: The Creation of a Self-Creating World* (New York: Crossroad, 1997), 162.

THE CONTRIBUTION OF PROCESS THEOLOGY TO EVOLUTIONARY THEOLOGY

The present-day significance of Whiteheadian process theology is also the product of a gradual development. Charles Hartshorne, an early disciple, brought Whitehead's ideas to the philosophy faculty of the University of Chicago, where he influenced theologians such as Daniel Day Williams, Schubert M. Ogden, and John B. Cobb Jr. The latter expanded this movement to Claremont Graduate University in California, where publications, events, and scholarly collaborations served to propagate Whitehead's ideas throughout the country and beyond. This movement, known as the Chicago School of process theology, peaked in the 1960s and 1970s.

Catholic and Orthodox theologians, however, were not quick to join. For instance, although in 1966 the Catholic theologian Eulalio Baltazar called for a switch to a "philosophy of process," his shallow critiques of Whitehead and Hegel showed that despite significant overlapping claims, Baltazar's true inspiration was Teilhard, not Whitehead or his followers.[76] Ewert Cousins's 1971 anthology first broke the ice between Catholicism and process theology, but his choice of excerpts made clear that the primary appeal of process theology for Catholic theologians still lay in its apparent resonance with Teilhard.[77] Eventually, in 1975, Cousins's student Robert B. Mellert finally promoted Whitehead's theology on its own terms.[78] Yet even today, Whitehead tends to find purchase among Catholic evolutionary theologians only by intermixture with Teilhard de Chardin. After all, Cousins was an expert on mysticism and spiritual theology. His influence helped to promote a more spiritual and less philosophically rigorous reading of Teilhard and Whitehead. The intermixture of Teilhard and process theology has become common among many evolutionary theologians, as seen for example in the works of Ilia Delio and Beatrice Bruteau.

Even when apparently operating in the background—peeking out amid footnotes and a few scattered, offhand references—process theology tends to have a significant impact on the claims of evolutionary theologians. When used as an add-on to Teilhard, for example, it goes hand in hand with a

---

76 Eulalio R. Baltazar, *Teilhard and the Supernatural* (Baltimore: Helicon, 1966), 102–103.
77 Cousins, *Process Theology*.
78 Robert B. Mellert, *What Is Process Theology?* (New York: Paulist, 1975).

tendency to foreground Teilhard's less common but more edgy statements: his emphasis on theodicy and the statistical necessity of evil, his critique of Aristotelian ontology, and his claim that God can only create by way of evolution. The more that process theology is able to assert itself, the more central such arguments become—especially inasmuch as they make direct and powerful claims not only about the evolving world, but about God's very own self.

While the full breadth of process theology's significance easily deserves a volume of its own, two aspects in particular are especially visible among contemporary writers. These include the subjection of divinity to theodicy and the restructuring of providence into the creative process.

God Defined by Theodicy

Many evolutionary theologians place theodicy—the problem of reconciling the existence of evil with the goodness of God—at the forefront of their considerations. Whitehead has played a major role in popularizing and shaping this emphasis, which continues to have a dramatic impact on theology today.

This important focus on theodicy did not begin with Whitehead, however. It is rather the product of a range of philosophical and religious of influences that have shaped modernity unto the present day. Leibniz coined the term "theodicy" in 1710, but John Milton foreshadowed its rise as a central concern of modern thought as early as 1667 when he opened his famous epic with the aim that: "I may assert th' Eternal Providence, / And justifie the wayes of God to men."[79] This program of divine justification met a serious roadblock in Kant's 1791 essay "On the Miscarriage of All Philosophical Trials in Theodicy."[80] Kant skillfully undermined the quest for theodicy, arguing that the issue could only be reconciled from the standpoint of practical reason (Kant's fill-in for "faith") and not upon any purely rational, speculative basis. This hardly put an end to theodicy's hold on the modern imagination, however, as Darwin's own ponderings illustrate. Yet it did put into question all-too-easy solutions such as those of Leibniz, who justified God's creation

---

79 John Milton, *Paradise Lost*, book 1, lines 25–26. This is the same in both the first and second editions of the text.

80 Immanuel Kant, "On the Miscarriage of All Philosophical Trials in Theodicy," in *Religion and Rational Theology*, ed. Allen W. Wood, trans. George Di Giovanni (Cambridge: Cambridge University Press, 1996), 30–31.

on the presumption that at least this represents "the best among all possible worlds."[81] Leibniz's view essentially justified the world *beforehand*; it must be good enough because, on principle, God always wills the best. In contrast, Hegel proposed a more dynamic "justification of the ways of God . . . so that the ill that is found in the world may be comprehended, and the thinking spirit be reconciled with evil."[82] For Hegel, God's creative action is justified by the process of the world as a whole. In a sense, the world *comes to be* the best inasmuch as its history represents the progressive coming-to-be of Spirit.

Importantly, prior to the modern era, God did not always require such justification. Many of the earliest portions of the Bible show little or no compunction about attributing violence and death to the Lord's own action (e.g., Exod 12:29). Later Jewish writers and early Christians added nuance to better insulate God from particular evils by, for example, developing the concept of Satan (1 John 3:8). Nevertheless, patristic and medieval thinkers tended to take God's righteousness as a given and to seek the origin of evil among human beings. While Augustine grappled with theodicy—for example, in his dispute with the Manichees—his responses took God's blamelessness for granted. In contrast and to varying degrees, modern approaches to theodicy followed Milton's angle of approach, putting God on trial in order to seek an acquittal. Ironically, the Reformers' emphasis on Christianity as the justification of sinners shifted to an emphasis on metaphysics as the justification of God.[83]

Whitehead's metaphysics essentially accepts this indictment of God but responds by denying God's ability to overcome evil. Since God requires the world and its creative process in order to become God, God is not really capable of eliminating evil outright. Rather, evil is an intrinsic and unavoidable element of a world created by way of process. In this sense, evil is in its essence the sense in which the world remains chaotic, disordered, and incomplete. Because process implies a movement from incomplete to complete,

---

81 Leibniz, *Theodicy*, 57.

82 Georg Wilhelm Friedrich Hegel, *Lectures on the Philosophy of History*, trans. Ruben Alvarado (Aalten, Netherlands: WordBridge, 2011), 14. Hegel's German reads *Rechtfertigung Gottes*, "justification of God." Johann Jakob Bodmer's 1732 German translation of *Paradise Lost* used the same term, which was traditionally associated with Martin Luther's doctrine of justification by faith: *Den Menschen die Wege Gottes rechtfertigen möge. Johann Miltons Verlust des Paradieses; ein Helden-Gedicht*, trans. Johann Jakob Bodmer, 1st ed. (Zurich: M. Rordorf, 1732), 2.

83 See Whitehead, *Religion*, 5, 73; Leibniz, *Theodicy*, 56–57.

evil remains a necessary component within the evolution of the good. Many of Whitehead's followers intensify these arguments by appealing to a sense of divine suffering, a notion somewhat less present in Whitehead than in Hegel.[84] Just as the world must suffer the existence of evil, so also does God suffer alongside the world.

As we shall see in chapter 5, Whitehead's emphasis on theodicy contributes significantly to many evolutionary theologians' reconceptualizations of original sin.

Providence Restructured

The question of theodicy connects intimately with the problem of divine providence. Does God intervene on behalf of God's creatures, and if so, to what extent? For Milton and his modern successors, to "justifie the wayes of God to men" inherently meant to "assert th' Eternal Providence."[85] To provide a rational justification for the existence of evil in the world appears, at first glance, to rescue Christians' belief in God's providential care. Nevertheless, the problem of divine providence has never been so simple. The centrality of theodicy within contemporary evolutionary theology has led to a proliferation of attempts to recue providence by gutting it. Process theology in particular places great importance on the concept of divine providence, but it requires that this concept undergo the same dramatic reconfiguration as the very concept of God.

In fact, the problem of providence runs deep within the history of philosophy. In a sense, Hegel and Whitehead reconfigure the concept of providence in order to overcome a difficulty that arose during the golden age of Greek philosophy.

In order to purify the gods of the whimsical and often malicious behavior attributed to them by Hellenic myth and religion, Plato emphasized providential care as one of their essential attributes.[86] Plato's gods were subordinate to the Ideas; the former existed and acted for the sake of the latter. The gods' providential care served to perpetuate the manifestation of the beauty, goodness, and order of the Ideas within the created world. In this way, Plato's

---

84   Whitehead, *Process and Reality*, 532 (351).
85   Milton, *Paradise Lost*, book 1, lines 25-26.
86   Gilson, *God and Philosophy*, 30; Plato, *Laws*, trans. R.G. Bury (Cambridge, MA: Harvard University Press, 1926), 352-77 (§§900-907).

notion of providence hinged upon the distinction between the gods as the subjective or personal ultimate and the Ideas as the objective or impersonal ultimate.[87] Fate, destiny, or providence were the product of will, and thus remained distinct from the kind of determination that stemmed from the impersonal, objective laws of nature.

Aristotle's philosophy effectively shattered Plato's concept of providence, however, because it collapsed the distinction, thrusting the divine into the role of philosophical first principle. Aristotle's Unmoved Mover is both objectively and subjectively ultimate. It performs an essential causal role, but without any personal notion of providence. Étienne Gilson explains, "The pure Act of the self-thinking thought eternally thinks of itself, but never of us."[88] Aristotle's approach reached a new degree of rational coherence because it recognized that a god could not be God while remaining subject to some external Idea or principle. However, it also made this God's relationship to the world anything but incidental. This God's sole contribution to the world's welfare consisted in being the original spark or initiator of all movement. Put another way, it became impossible to discern God's free subjectivity within the objective and necessary machinations of nature's laws. The pantheism of the Stoics (ca. 300 BCE–180 CE) repudiated key elements of Aristotle's view but effectively exasperated this collapse. Rejecting the aloofness of Aristotle's God, they instead saw divinity as an inner vital and creative principle within all things. In this way, God's immanent action within the world became indistinguishable from the blind determinism of natural laws.[89]

Like the Neoplatonists before him, Hegel takes rescuing divine providence as a central philosophical goal.[90] Medieval theologians understood providence as God's universal care for all beings, and with regard to the human world

---

87 See Gilson, *God and Philosophy*, 21–22.

88 Gilson, 33. Contemporary critics often accuse scholastic theology of adopting Aristotle's disinterested God because of its appropriation of the concept of God as Prime Mover. In reality, however, the robust scholastic understanding of divine providence testifies to the fact that the scholastics subjected Aristotle's contributions to standards set by the Bible and Christian tradition. See Joseph Ratzinger, "The God of Faith and the God of the Philosophers," in *Introduction to Christianity*, rev. ed. (San Francisco: Ignatius, 2004), 137–50.

89 Note, however, that the Stoics identified god with nature only in a subtle and qualified sense. See Michael J. White, "Stoic Natural Philosophy (Physics and Cosmology)," in *The Cambridge Companion to the Stoics*, ed. Brad Inwood (Cambridge: Cambridge University Press, 2003), 137.

90 Hegel, *Lectures on the Philosophy of History*, 12–13. See Plotinus, *Enneads*, 2.9.9; 3.2–3.

they tended to see this providence principally in terms of God's loving care for persons. God's care for individuals was the model for understanding God's providence toward peoples, nations, and the Church. Hegel, in contrast, sees providence as a structural element of history as such. If God cares for individuals, this is only a secondary and indirect manifestation of God's more fundamental role in shaping the history of the world through human civilization. Hegel annuls any distinction between salvation history and secular history.[91] Divine providence means that history as such is the teleological process whereby the future realizes itself within the back and forth churnings of the present. Providence is not as an extrinsic, extra- or suprahistorical intervention that guarantees the happiness of persons but rather as an intrinsic element of history as such.

Whitehead's take on history and providence closely resembles that of Hegel. For both, history is intimately governed by reason, and reason is in that sense providential. "Reason is the special embodiment in us of the disciplined counter-agency which saves the world."[92] What this means in particular is that history is the gradual development of rational order through the persuasive inculcation of Ideas.[93] Like Plato's gods, the God of Whitehead is inherently providential precisely because this God—in terms of God's primordial nature—serves as a medium for the translation of Ideas into real life. Each actual occasion is a "synthesis of the ideal with the real," which is accomplished through the influence of "the Divine Eros as the active entertainment of all ideals, with the urge to their finite realization."[94] By being above change, God's primordial nature establishes consistency and order in the overall process of the world.[95]

In this way, both Whitehead and Hegel see the historical development of human civilization as meaningful, providential, but not deterministic. Both see history as an interplay of freedom and necessity, a process by which freedom becomes the ultimate meaning of the world.[96] In effect, world history is both rationally structured and open to free self-determination. In retrospect, history's events make philosophical sense, but they are not thereby utterly predictable or predetermined.[97]

---

91 Cyril O'Regan, *The Heterodox Hegel* (Albany, NY: SUNY Press, 1994), 274.
92 Whitehead, *Function*, 28.
93 Whitehead, *Adventures*, 32–33, 72–73.
94 Whitehead, *Adventures*, 276.
95 Whitehead, *Religion*, 85–86.
96 Hegel, *Lectures on the Philosophy of History*, 24.
97 Whitehead, *Religion*, 82.

As Whitehead explains it, divine providence operates not by forcing history to move in this or that way but rather as a persuasive "lure for feeling" or varieties of experience. The primordial nature of God serves to entice reality toward the ever deeper embodiment of the Ideas in the formation of actual occasions. Whitehead's God is thus far more involved in the world than an Aristotelian First Mover or a deist God who only initiates the flow of history. Nevertheless, Whitehead's God does not undertake to care for specific individuals and does not intervene beyond this function as persuasive lure.

In view of the above, it is clear that for Whitehead, no real distinction stands between creation and providence. God's creative agency is not a once-and-done event of the past but rather the ongoing process by which God provides a creative lure for existence.[98] God is integral for the coming-to-be of every actual occasion, not because God is the efficient cause or even one cause among others, but rather because God effectively imports meaning into the process of concrescence by stirring up the urge toward the realization of the Ideas.

Yet by making God merely one aspect among others in the process of concrescence, process theology relativizes God's creative role. "The world is self-creative," writes Whitehead.[99] God's role is only to be "the lure for feeling, the eternal urge for desire."[100] God is the eternal object *par excellence*, which stirs up the subjective aim in every process of concrescence. Thus God's participatory creation in no way supplants either the ordinary causality of the world or the free self-realization of actual occasions. Divine creativity is merely another way of looking at the process of creaturely self-creation. Influenced by Whitehead's views, Philip Hefner has popularized the description of humans as "created co-creators."[101]

Whitehead, Teilhard, and Beyond

These contributions add up to a substantial shift in evolutionary theology, which has helped to form more distinct schools of thought within it. Despite

---

98 Whitehead, *Religion*, 91; *Process and Reality*, 526 (346).
99 Mellert, *What Is Process Theology?*, 130.
100 Whitehead, *Process and Reality*, 522 (344).
101 Philip J. Hefner, *The Human Factor: Evolution, Culture, and Religion* (Minneapolis, MN: Fortress Press, 1993), 27.

many points of overlap and the potential for compatibility, there are important differences between Teilhardian and Whiteheadian forms of evolutionary theology.

For example, process-based theology need not be as teleological as that of Teilhard de Chardin. Put another way, the freedom of the future tends to carry a more explicit and fundamental role within process-based forms of evolutionary theology. Teilhard's view is not deterministic, but he does see the future as radically pre-embodied in Jesus Christ. This leads him to speak of the future with a sort of eschatological confidence. This also means that Teilhard's theology is more apocalyptic; it sees the eschatological future of the world as already gradually revealing itself in and through the world's ongoing evolutionary history. For Whitehead, in contrast, the future serves not so much as the real goal of history but rather as its horizon of open possibilities. In a sense, process theology sees the world's endless creative journey of self-becoming as so meaningful in itself that any ultimate fulfillment, by implying a cessation of this process, could only disappoint. The goal of history in Whitehead's view is not some specific and predetermined outcome but rather a maximalization of the ultimate enjoyment of actual existence.[102]

Perhaps nothing illustrates this difference better than the meaning of Christ. Process theology sees evolution as a fundamental aspect of God and reality even prior to and apart from Christ. As the incarnation of God, Jesus Christ is perhaps the best and most perfect symbol of the God–world relationship, but he himself does not actually constitute that relationship.[103] Teilhard, on the other hand, depends upon a strongly Catholic Christocentrism, such that Christ is always more than a mere symbol of God. In fact, the evolution of the world can only really function and achieve its meaning in and through this Christ. While incarnation does represent something true about the broader God–world relationship, this is only because it is perfectly realized in the one God-man, Jesus Christ.

In sum, process theology is responsible for many significant trends within contemporary theology and within evolutionary theology in particular. It has become a major part of the theological conversation about God, the world, and the relation between these two.

---

102 See Whitehead, *Religion*, 87.
103 Cousins, *Process Theology*, 31–32.

EVOLUTIONARY THEOLOGY

Having examined the two most significant fountainheads of evolutionary theology, we turn now to consider some key dogmatic issues and how they have spurred important theological considerations. This will put us in a position to understand the significance of evolutionary theology for the doctrine of original sin and to explore evolutionary possibilities for the future.

CHAPTER FOUR

Dialoguing between Doctrine and Science

As we have seen, Whitehead's theology shakes the very foundations of Christian tradition by portraying the evolving world as the very process of God's own coming-to-be. Yet not all evolutionary theology is quite so daring. For key figures such as Karl Rahner (1904–84), the goal is not to create a new theology based entirely upon the idea of process or evolutionism but rather to understand how traditional ideas, frameworks, and viewpoints can fruitfully dialogue with the science of human evolution.

With such an integrative goal, it is no coincidence that Rahner, like Teilhard and many others, speaks out of the Catholic tradition. Catholicism leans in this direction in many ways. For example, it has long recognized the possibility of reading the creation stories of Genesis more figuratively than literally; it emphasizes the unity and rationality of the world; and it favors an approach that sees the world as fundamentally revelatory of a benevolent God. Addressing the Pontifical Academy of Science in 1951, Pope Pius XII stated, "By your research ... you preach at the same time, in language of figures, formulas, and discoveries, the unspeakable harmony of the world of an all-wise God.... [T]rue science discovers God in an ever-increasing degree."[1]

This affirmative stance toward science *in principle* does not, however, preclude Catholicism from exhibiting at the same time a tendency toward intellectual conservatism. The Church's official responses to Darwinian evolution were slow, complex, and largely negative up until 1950. Such a revolutionary

---

1 Albert Schlitzer, "The Position of Modern Theology on the Evolution of Man," *Laval théologique et philosophique* 8, no. 2 (1952): 212.

idea as evolution had easily appeared suspect in light of the crisis of Catholic modernism at the turn of the century.

Nevertheless, the Church's early stance toward the theory of evolution was more subtle and complex than even most Catholics at the time recognized. Magisterial pronouncements on the issue were few, limited in scope, and sometimes open to interpretation. Despite important theological reservations, the Holy See did little to nothing to suppress the theory of evolution. In the case of Teilhard de Chardin, his Jesuit superiors prevented him from publishing the bulk of his theological writings prior to his death in 1955 because of his problematic assertions on original sin and evil. Yet despite these issues, neither the Holy See nor the Society of Jesus took any efforts to prevent Teilhard's posthumous publication.[2] Importantly, Teilhard's theology was suspect, but never his *science*. Since the late nineteenth century, many Catholics privately approached evolution—even human evolution—as an open question. By the 1940s and 50s, other Catholic paleontologists such as Piero Leonardi, Bermudo Meléndez, and Miquel Crusafont were already engaged in spreading teleological (and thus implicitly theological) forms of the doctrine of evolution prior to Teilhard's broad publication.[3]

Rather than attempting to reconcile evolution with a literal reading of the creation stories of Genesis, the Catholic Church's official engagement with the theory of evolution has hinged primarily upon key dogmatic questions concerning the nature of creation, the relationship between God and the world, the dignity and spiritual nature of humankind, and the reality and inheritance of original sin. These questions have helped to shape evolutionary theology as a result. In other words, the complex relationship between Catholicism and evolution has become a structuring element; the key dogmatic questions posed by Catholic thinkers continue to impact many forms of evolutionary theology today.

---

2 While his conflicts with the ecclesiastical authorities were no doubt difficult and taxing, it is not uncommon for later accounts to exaggerate the significance and extent of the penalties imposed upon him. See Donald Wayne Viney, "Teilhard: *Le Philosophe malgré l'Église*," in *Rediscovering Teilhard's Fire*, ed. Kathleen Duffy (Philadelphia: Saint Joseph's University Press, 2010), 69–76. Anyone who even critiqued Teilhard's ideas during his lifetime—whether they be his Jesuit superiors or independent theologians—is equated with "Rome" or "the Church" in order to paint a politicized image of Teilhard versus the Holy See.

3 Francisco Blázquez Paniagua, "La recepción del darwinismo en la universidad española (1939–1999)," *Anuario de Historia de la Iglesia* 18 (2009): 57, 60–61.

Hence, even without entering into the contentious debates between science and religion in regard to evolution—debates that involve Catholics as well as Protestants—a terse examination of the impact of the Catholic Church's magisterial response to evolution will provide key insight into the trajectory of evolutionary theology during the latter half of the twentieth century. This history reached a pivotal moment in 1950 when Pope Pius XII's encyclical letter *Humani generis* posed the very dogmatic questions that would frame much of the discourse of evolutionary theology for decades to come. An analysis of key answers to these questions by Karl Rahner will show how they set the stage for present-day theologians.

FRAMING THE GROUNDWORK: *HUMANI GENERIS*

We cannot in this short space undertake a full historical treatment of the various Catholic responses to evolution since Wallace and Darwin. While one might briefly summarize the Church's attitude as reserved or defensive up until the Second Vatican Council (1962–65), such an assessment would do justice neither to the diversity of voices within the Church—lay and ordained—nor to the careful and often political subtlety of magisterial responses. In view of this, without attempting to draw a complete picture, the most effective approach to understanding the history of Catholicism's engagement with Darwinism is to view the key questions and themes through the lens of *Humani generis*. This groundbreaking encyclical both summarized what was most important in the debate leading up to that point and provoked new attempts to reconcile evolutionary theory with the fundaments of Catholic theology.

While most nineteenth-century Catholic bishops and theologians did not rush to embrace the idea of natural selection, neither did the Church definitively reject it outright. Catholics enthusiastically entered into the debate on both sides. At the time, St. George Jackson Mivart (1827–1900)—perhaps the most significant early Catholic supporter of evolution—felt entirely justified in claiming that even "the strictest Ultramontane Catholics are perfectly free to hold the doctrine of evolution."[4]

---

4 St. George Jackson Mivart, "Modern Catholics and Scientific Freedom," *The Nineteenth Century* 18 (1885): 34. Note that "St. George" is literally his personal name. He is not a canonized saint.

Nevertheless, a shifting tide led Catholic intellectuals to adopt a prevailing sense of caution or even suspicion with regard to this doctrine. From the start of the twentieth century up until about 1950, Catholic textbooks widely considered evolution to be, if not dogmatically excluded, then at least "rash" and "unsafe."[5] Many did not consider faithful Catholics to be free to hold the theory. This situation owed a lot to public perception of the cases of two figures who were widely seen as disciples of Mivart: Dalmace Leroy (1828–1905) and John Zahm (1851–1921). In 1902, *La Civiltà Cattolica*, a Roman Jesuit journal with close ties to the Holy See, published an essay by Salvatore Brandi that gave a strong impression that Leroy and Zahm had been made to retract their views by a formal inquest of the Holy Office, the doctrinal watchdog of the Holy See. Because such proceedings would be private and not readily verifiable, Brandi's assertions were difficult to disprove and became commonly accepted. The names Leroy and Zahm became watchwords for the inadmissibility of evolution.

In truth, the supposed condemnations of Leroy and Zahm ceased to hold water by the mid-twentieth century owing in large part to the fact that they never actually occurred.[6] The truth is far more subtle. Leroy and Zahm were never censured by the Holy Office. Their respective books were, however, under threat of being put on the *Index of Prohibited Books* by the Congregation of the Index. Such indexing does not hold the same doctrinal weight, nor does it imply that the author's views are altogether inadmissible. Nevertheless, even this did not happen; the works were never actually added to the *Index*. Owing to the secrecy of the proceedings—a customary practice protecting the privacy of those who were investigated—the actual outcome of these cases was shrouded by rumor. Brandi's misrepresentation of the facts became sufficient evidence to cast a public shadow over evolution in Catholic thought for many decades. Nevertheless, the rampant public misperception generated by Brandi's assertions holds no magisterial authority. Despite widespread belief to the contrary, when Pius XII penned *Humani generis*, the admissibility of evolution was still technically and officially an open question.

For Pius, the fundamental unity of truth necessitates that the Church come to terms with scientific ideas. In his view, if science seems to contradict the

---

5 Schlitzer, "Position of Modern Theology," 210.
6 Mariano Artigas, Thomas F. Glick, and Rafael A. Martínez, *Negotiating Darwin: The Vatican Confronts Evolution, 1877–1902* (Baltimore: Johns Hopkins University Press, 2006), 16–17, 29.

Church's teachings, then it is only because people have misunderstood the evidence or have accepted mere unproven hypotheses as genuine fact.[7] A genuine rapprochement between theology and science is not only possible but praiseworthy, and yet it requires a careful weighing of evidence in a way that gives due deference both to the rigor of modern science and to the authority of divine revelation.

Given these points, Pius sees the issue of evolution as an open question that demands further consideration by scientists and theologians alike. "The Teaching Authority of the Church," he writes, "does not forbid that, in conformity with the present state of human sciences and sacred theology, research and discussions, on the part of men experienced in both fields, take place with regard to the doctrine of evolution."[8] This openness is not without specific reservations, however. Pius lays out two key issues that might stand in the way. These two issues represent theological, dogmatic obligations that may not be compromised by the acceptance of biological evolution.

First, such evolution cannot apply with regard to the human soul in the same way as to the human body. While the origin of the soul had been an open question during a period in early Church, contemporary Catholic theology considers it certain that one's soul is not inherited from one's parents. Rather, each soul is directly created by God. The human body can be the product of gradual development, even by natural selection, but the soul is neither something material nor a merely quantitative improvement of an earlier form. Spirit, properly understood, cannot be the product of biological development.

Such a clear restriction may seem to admit no freedom in regard to the evolution of spirit, but many Catholic thinkers before and after the encyclical found ways to creatively expand the bounds of orthodoxy in this regard. As we shall see, Karl Rahner argued that spirit could be seen as related to the evolution of matter without compromising the belief in the immediate creation of the soul.

Second, the science of evolution may run into serious theological pitfalls when embracing the theory of *polygenism*. Polygenism, like this second issue as a whole, is complex and multifaceted. In its typical though imprecise usage, polygenism refers to any position that argues that humans descend from multiple different origins. For some polygenists, this means more categorically that there was never an original Adam and Eve; humans could not have descended

---

7 Vatican I, *Dei Filius*, ch. 4.
8 Pius XII, *Humani generis* (New York: Paulist, 1950), §36.

from a single original couple. Nevertheless, polygenism often refers to what is more properly termed *polyphyletism*, the theory that multiple, distinct prehuman populations (i.e., of *Homo erectus*) around the world evolved separately into modern-day humans (i.e., *Homo sapiens*). Still, many polygenists—among them Teilhard de Chardin—believed that even this second form of polygenism or polyphyletism also contradicted the existence of Adam and Eve.

Despite this recurring denial of Adam and Eve, for *Humani generis* the key problem of polygenism is not, strictly speaking, that it contradicts the letter of Scripture. Rather, by disputing the genetic unity of the human race, polygenism risks undermining the basis of the doctrine of original sin. The doctrine states that all humans have inherited culpability and a sinful inclination from the sin of our first ancestors. This inheritance is made possible by the fundamental unity of the human race, and for Catholic tradition that unity is first and foremost biological. We are all biologically related.

Pius's language here is extremely careful; he does not want to overstate his position and ban too rashly the idea of human evolution as a whole. Nevertheless, in regard to original sin, the theory of evolution presents no minor problem. He writes:

> When, however, there is question of another conjectural opinion, namely polygenism, the children of the Church by no means enjoy such liberty. For the faithful cannot embrace that opinion which maintains that either after Adam there existed on this earth true men who did not take their origin through natural generation from him as from the first parent of all, or that Adam represents a certain number of first parents. Now it is in no way apparent how such an opinion can be reconciled with that which the sources of revealed truth and the documents of the Teaching Authority of the Church propose with regard to original sin, which proceeds from a sin actually committed by an individual Adam and which, through generation, is passed on to all and is in everyone as his own.[9]

At the time, theologians were divided on exactly how severe Pius intended this critique to be. Does "it is in no way apparent" mean that it is impossible to reconcile theology with polygenism? Or does it hold out hope that a solution might become apparent in the future? Does the critique of polygenism include polyphyletism, or does it apply only to the matter of Adam and Eve?

---

9 Pius XII, §37.

One way or another, Pius shows that the Magisterium's role is not to decide for or against the science of evolution, but only to indicate to what extent this science may be "safe" for consideration. The question becomes, can a faithful Catholic ascribe to the theory of evolution without a danger of compromising the tenets of faith? This question will set the stage for the key approach of Karl Rahner and make possible the future expansion of evolutionary theology within Catholic thought.

These two principal issues highlighted by *Humani generis* brought about new fervor among Catholic theologians who sought to identify solutions that could reconcile human evolution with these key dogmatic commitments. Far from limiting or restricting evolutionary theology, they formed the vital questions that impelled it forward. First, how can and to what extent can we understand the human body as governed by evolution without making the soul a mere byproduct of the body's development? Second, how can we maintain the hereditary unity of the human race so as not to contradict the traditional doctrine of original sin? We turn now to see how these two questions shape the ongoing history of evolutionary theology.

THE FIRST QUESTION: EVOLUTION AND THE SOUL

At first glance, *Humani generis* may seem to settle the problem of human evolution by allowing the body to be the product of biological development while retaining the soul as a miraculous and direct creation of God. The body can belong to biology so long as the soul remains the property of theology. However, as Karl Rahner points out, the Church's understanding of the unity of the human person forbids any such dualistic solution. Just as one cannot simply isolate science and theology into completely distinct, separate, and autonomous spheres, so also does the human person evade any simplistic dissection. For theology to speak about the soul inherently implies also a statement about the body. Likewise, what science teaches us about the body must also have real consequences for the human person as a whole.[10]

Pius XII's insistence on maintaining the doctrine of the immediate creation of the soul must be read in light of the fundamental unity of the human person. Science is especially competent to speak in regard to the human body. However,

---

10 Karl Rahner, *Hominisation: The Evolutionary Origin of Man as a Theological Problem*, trans. W.J. O'Hara (New York: Herder and Herder, 1965), 18.

since the proper meaning of the body is realized only in its union with the soul, even a scientific understanding of the body contributes to the full theological meaning of the human person. Pius's refusal to simply hand over the soul is no mere act of stubborn dogmatism. Rather, it stands as an invitation to explore the truth of the human person as a whole—body and soul—through a fruitful and cooperative dialogue between science and theology.

Rahner took up this invitation shortly after the 1950 publication of *Humani generis* and continued to develop his views for the next twenty years. His 1961 book *Hominisation*—an expansion of an earlier essay—treats at length the problem of the soul's immediate creation in relation to human evolution. Beginning from a Thomistic standpoint, Rahner argues that it is possible to maintain the immediate creation of the soul while also, in a certain sense, recognizing a certain intimate connection between the development of the body and the creation of the soul. The soul cannot be reduced to a mere product of natural development.[11] It did not evolve as a purely natural improvement of humanity's purely biological capabilities. Nevertheless, the creation of the soul is premised upon the evolutionary history of the body in such a way that bodily matter can be seen as the authentic "pre-history" of spirit.[12]

Rahner's approach was not altogether new. Similar treatments of the relationship between spirit and matter can be found in the works of Mivart and Teilhard de Chardin, for example.[13] Yet Rahner's theological acumen and strong Thomistic basis added to his influence upon subsequent Catholic thought. A younger Joseph Ratzinger (Benedict XVI) foreshadowed the growth of this influence in a 1968 radio talk. Borrowing Rahner's assertion that matter is "the pre-history of spirit," Ratzinger explains:

> It is clear that spirit is not a random product of material developments, but rather that matter signifies a moment in the history of spirit. This, however, is just another way of saying that spirit is created and not the mere product of development, even though it comes to light by way of development.[14]

---

11 See Christoph von Schönborn, *Man, the Image of God: The Creation of Man as Good News*, trans. Henry Taylor and Michael J. Miller (San Francisco: Ignatius Press, 2011), 64.

12 Rahner, *Hominisation*, 63.

13 St. George Jackson Mivart, *On the Genesis of Species*, 2nd ed. (London: Macmillan, 1871), 325–31; Teilhard de Chardin, *Human Energy*, 43.

14 Horn and Wiedenhofer, *Creation and Evolution*, 14.

Rahner's perspective sought to allow for evolution not by surrendering certain points of doctrine, but rather by rediscovering and expressing anew the real meaning of vital doctrines in a way that revealed their inner compatibility with new forms of thought.

After all, the Church's belief in the immediate creation of the soul did not originate as a challenge to biological evolution. Rather, it developed during the Church's first few centuries out of its conviction that each human person bears a unique, direct, and inescapable connection to the Creator. No one is an accident. Each individual is loved by God in such a direct way that God wills that person's existence as well as happiness. As Ratzinger points out, the doctrine of creation does not place Adam in an exclusive position; instead, it speaks about the real relationship each of us has with God. "The faith declares no more about the first man than it does about each one of us," he says, "and, conversely, it declares no less about us than it does about the first man."[15]

In order to better understand this, we will look closer at Rahner's treatment of the evolutionary relationship between spirit and matter. First, this requires some clarification of the terminology involved.

The Terminology: Spirit, Matter, and Form

The terms "spirit" and "matter" have had a long, complex, and multifaceted history. They have not always meant the same thing in every discourse. Even today, these terms can vary in meaning depending on context. Accordingly, they can be gravely misleading if not read in light of the philosophical traditions involved.[16]

Rahner's particular use of these terms is rooted in the relatively stable ground of Thomistic tradition. Even so, Rahner's references to "matter" and "spirit" are not simply interchangeable with "body" and "soul." This brief introduction allows too little space to adequately expound upon the full significance of Rahner's terminology. However, we should at least clarify four particular pairs that are central to understanding his argument: scientific matter as opposed to

---

15 Horn and Wiedenhofer, 14–15.

16 For brief definitions, see Rahner, *Dictionary of Theology* (hereafter DT), trans. Richard Strachan, 2nd ed. (New York: Crossroad, 1985), s.vv. "Form", "Matter", "Spirit." See also Theodore Kepes Jr., "Toward a Unified Vision: The Integration of Christian Theology and Evolution in Karl Rahner's Understanding of Matter and Spirit," *Philosophy and Theology* 20, no. 1–2 (2008): 269–90.

energy, Platonic matter as opposed to spirit, Aristotelian matter as opposed to form, and Thomistic body as opposed to soul. Because of the interdependence of each term upon the other, they naturally function as dualities. For Christian tradition they are not, however, dualisms. When properly understood, for example, matter obtains its meaning in view of spirit, but matter is not something altogether opposed to or in conflict with something called spirit.

(1) *Scientific matter and energy.* In the first place, the prevalence of scientific discourse in the modern world has dispersed a concept of matter that, while historically rooted in earlier philosophical ideas, is nonetheless altogether distinct from the Thomistic concept. Especially since Albert Einstein's 1905 publication of his special theory of relativity, scientific matter can only be understood in relation to physical energy. For science, matter is a quantity with measurable mass and volume. It is a localized, concentrated form of energy; volatile by its very nature, it carries always the potential to be transformed back into energy in a fluid sense. Energy, properly understood, is not a substance flowing through electrical wires or other materials, but rather the potential for some physical action or transformation. A bouncing ball has energy both in the form of mass and in the form of bouncing.

Because of the convertibility of scientific matter and energy, both of these count as matter from the standpoint of Thomistic philosophy. Whereas science uses the term matter to refer to one aspect of reality—that which has measurable mass and volume—Thomistic philosophy uses this term to refer to sensible reality as a whole. All objects of our concrete experience are, in fact, material—even the emptiness of space. Even emptiness is a "something," which would not exist were it not for God's gratuitous creative action. Hence, the scientific distinction between matter and energy has no relevance for what theology calls the soul. To say that the soul is immaterial is not to say that it is some sort of energy, as though it were a static electrical charge or magnetic field. The soul is neither mass nor energy nor anything else that enters into the direct experience of our five senses.

(2) *Platonic matter and spirit.* Although Thomas is famous for his appropriation of Aristotle, it is important to note that his views are still shaped also by Platonism. Like the majority of Christian theologians, Thomas does reject Plato's notorious, dualistic opposition between matter and spirit. However, this is not to say that he rejects Platonism as a whole.

Christianity's use of matter and spirit derives from Platonism's need to differentiate tangible, sensible reality from intellectual, conceptual reality.

Matter characterizes the world that we encounter with our bodies and their senses. Spirit, by contrast, is that which remains irreducible to such bodily encounters. Since knowledge requires a likeness between the knower and the known, the character of the human person as spirit enables one to know immaterial realities—specifically, the transcendent Ideas or forms after which the material world is patterned. In a sense, this means that spirit is another way of looking at the human person. However, it also means that the person cannot be reduced to matter alone.

(3) *Aristotelian matter and form.* Aristotle rejected Plato's belief that Ideas exist on a transcendent plane independent of matter. Instead, he focused on the sense in which the intellect naturally abstracts the forms or essences of beings through contact with the material world. When we look at a table, we naturally compare it to a concept of "tableness." If someone points to a stick and calls it a table, we immediately compare the form of the stick to the form of tableness and reply, "That is no table." Materially speaking, both the stick and the table may be made of wood, but the form is what makes something a table rather than a mere stack of lumber. A form is thus an intelligible image that defines the shape, characteristics, meaning, and purpose of an object.

By comparison, matter is that which stands under the form. If you strip away the form of a table, you have only the matter of wood. Nevertheless and importantly, matter is here still very much an intellectual concept. It is not something that we encounter in raw experience. Every object of experience is already a unity of form and matter. This is because no matter what the object may be, one can still abstract a form that stands in contrast to its matter. Take for example our stack of lumber. Its matter is wood, and this takes the form of boards. Looking at the wood alone, this too is a form–matter unity. Materially, it is a complex concoction of cellulose and tannins; formally, these are structured into a matrix of cells and channels. When we mentally strip away this form, we are faced with the form of chemicals, then of atoms, then of subatomic particles. So long as we can think of an object in terms of its shape, characteristics, meaning, or purpose, that object must have a form. In relation to this, the materiality of the object is its raw potential for form. In regard to the table, the lumber has the potential to be made into a table, while the form is the actual shaping that makes this lumber a table.

(4) *Thomistic body and soul.* Thomas appropriated a key link between Aristotle's matter-form duality and the Christian distinction between matter and spirit. Pushing back against Plato's view that the soul used the body

as an instrument—like a charioteer driving a chariot[17]—Aristotle argued instead that the soul is the form of the body. The soul is what makes the body more than a lump of flesh; it is not only its life but also the characteristics and personhood of the individual. Thomas adopted and modified Aristotle's view. Understanding the soul as the form of the body helps to emphasize the fundamental unity of the two in a way that accords well with early Christianity's emphasis on bodily resurrection. Nevertheless, from as early as the second century, Christians have believed that at death the soul continues without the body and awaits the final resurrection. Thomas had to make room for this idea within his Aristotelian framework. Thus he argued that the soul is not an ordinary form. When a table is destroyed, its "tableness" does not continue to exist. Not so the soul. If the soul is capable of existing without the body while it waits for the resurrection, Thomas averred, then it must be a special *subsistent form*—a unique kind of form with a more flexible relation to matter.[18] Properly understood, the soul is not meant to be without the body. However, the death and destruction of the body cannot effect the nonexistence of the soul.

All four of these dualities play into Rahner's Thomistic standpoint. His use of matter and spirit, for example, does not imply a return to a strictly Platonic understanding of these terms, but it does imply a Christian understanding that has been influenced by the history of Platonism. If the human person is both matter and spirit, then we belong to two worlds. Materiality represents our intimate relation with the rest of the natural world. Spirituality represents the way in which we are irreducible to the natural world and, at the same time, able to commune with that which is beyond the world: God. While God is not spirit in the same sense as we are, God is better understood as spirit because it is this transcendent dimension of humanity that bears a closer analogy to the utterly transcendent reality that we call God. Nevertheless, a full understanding of humanity cannot consider only one of these two dimensions in isolation from the other. Accordingly, the spiritual dimension of humanity shows forth its full meaning only in light of humanity's material dimension. In fact, our communion with God is not a function of the spirit *as apart from* matter, but rather of the spirit *in relation to* matter.

---

17 Plato, *Phaedrus*, trans. Robin Waterfield (Oxford: Oxford University Press, 2002), 28.
18 Thomas Aquinas, *Summa theologiae* I, q. 75, a. 2; q. 76, a. 1 ad 5.

The Evolution of Matter toward Spirit

With these definitions in mind, it becomes impossible to think of spirit as merely a higher kind of material reality. Minerals naturally evolve out of thermonuclear reactions within stars; landscapes evolve from the movement of plates, water, winds, and animals; animals evolve from the alteration of genetic codes and environmental influences. Spirit, in contrast, is not a material reality and thus cannot evolve on the basis of any natural occurrences. Since spirit is by definition that which transcends mere material reality, no recombination of matter could ever result in the advent of spirit. They belong to fundamentally different orders, just as no combination of smells could ever produce the color blue. In this way, without completely redefining what Christianity means by "spirit" and "soul," the evolution of the body could never produce the human soul. Pius's exclusion of the soul from evolution follows logically and unavoidably from a theological understanding of the human person. From such a standpoint, the claim that evolution produces the soul would be not simply heretical but nonsensical. To use a rough and not unproblematic analogy, it would be like claiming that a television set *produces* the series on its screen.

Rahner's task, therefore, is not to overturn Pius's exclusion but rather to understand evolution in light of the real meaning of matter and spirit. After all, while spirit is indeed defined by its transcendence or otherness relative to body or matter, this definition also relates the two in such an intimate way that one really only makes sense in view of the other. In other words, the evolution of the human body cannot be fully understood apart from spirit. If the very meaning of our bodily matter is tied to the meaning of spirit, and if moreover the very form of the body is the soul, then any development of the body has immediate and unavoidable spiritual consequences.

Rahner is well aware that the operation of what we call the soul is both made possible and constrained by the material, biological, and chemical situation of the human brain. The existence of the soul implies that human biology has reached a level of complexity that allows for such self-transcendence. Mere chemical imbalances, whether natural or drug-induced, can dramatically affect our ability to express love, anger, forgiveness, and regret. Hence while what we call spirit is not reducible to mere emotions or attitudes, or even the chemical and electrical signals that underlie them, neither is the experience of oneself as spirit simply distinguishable from the biochemical processes by which our spiritual existence is materialized in the body.

The advent of the human spirit, therefore, must be both precedented and freely given (unexacted) at one and the same time. It is precedented: looking backward, the advent of the soul makes perfect sense in light of the evolutionary development of the body and its brain. In fact, our spiritual existence is premised not only on the body's growth in intellectual power but also the development of its emotional, social, and imaginative capabilities. Thus, our physical, biological capacities are in some sense prerequisites or precedents for our existence as embodied, spiritual beings. Yet the spirit is also unexacted: the advent of the soul comes about freely or even miraculously, because no amount of biological development can forcefully create an embodied, spiritual being. Although our spiritual capacity appears to follow naturally from the progressive development of the human brain, no amount of processing power can simply produce what we mean by spirit. No sufficiently fast computer must simply be *ipso facto* a spirit; no clever ape must by its sheer cleverness be what we call a person.

In the language of Søren Kierkegaard, the advent of spirit is a *qualitative* shift rather than a *quantitative* addition. The qualitative leap toward spirit is a change in kind rather than intensity, and yet it is in some sense precedented (though it cannot be forced) by the body's quantitative development—its growth in intensity in regard to its mental, emotional, social, and imaginative faculties. Only something already external and transcendent can bring about a true qualitative shift, and yet when it comes about, it makes perfect sense in light of the foregoing quantitative buildup toward spirit on the part of the body.

Without reducing the extraordinary difference between matter and spirit, Rahner proceeds by showing that such a qualitative leap is actually, in a sense, quite mundane. In fact, the qualitative leap represents the very logic behind Aristotle's form–matter union. Matter is in effect a kind of (quantitative) buildup toward form. Lumber, for example, has a very real and intense potential for becoming a table. Nevertheless, without the actual qualitative addition of an extrinsic form, that lumber will never actually become a table. The movement from mere potential to act can only occur, for example, when a carpenter makes the conscious decision to infuse the lumber with a table's form. Nevertheless, though it could not bring about a table without external help, the underlying material—the lumber—remains a vital and indispensable aspect of the table itself. Just as matter requires form, so also does form require matter in its own way.

Aristotle refers to the process by which form becomes embodied in matter as *generation*. In itself, this term is somewhat vague; it could refer to events as disparate from one another as the procreation of a fox kit and the coining of a new ad slogan. In short, there is nothing explicitly biological about Aristotelian generation. Nevertheless, Rahner sees in the term a way to conceptualize both the evolution of the body and the advent of spirit.[19] Without disparaging the scientific theory of natural selection or any other biological mechanism, the philosophical difference between a parent species and its descendant can be thought of in terms of form and matter. The parent species becomes the material basis of a new and relatively transcendent form. For example, beginning roughly fifty million years ago, the ancestor of the modern horse underwent slow changes that, through innumerable steps, begat the equine that we know today. Each step can be seen as the addition of a new form. Primitive eohippus was the matter; it took on a new form that modified its teeth, and this resulted in a new form–matter unity called orohippus.

In this way, Rahner argues that the evolutionary development of creatures points to a certain creative possibility within creation itself. Like Bergson, Teilhard, and others, Rahner does not want to reduce evolution to a mere unfolding of preexisting possibilities. If it generates something authentically new, however, then this newness must be made possible by God's transcendent causal relationship to the world. He suggests that Aristotelian metaphysics makes this possible by asserting that a form can arise from within the underlying matter.[20] Eohippus carried within itself the potential for a real self-transcendence; when actualized, this resulted in the evolution of orohippus.

In fact, if biological evolution can be understood in terms of form and matter, then it can be seen as already implicitly more than a *material* occurrence. Most Christian theologians cannot get around the belief that biological evolution is driven by some kind of purpose or teleology, and this form–matter analysis provides a way of understanding the teleology (provided by form) as intrinsically bound up with evolution's material progress. If the evolution of the three-toed sloth represents the arrival of an authentically new form, then it is not only inherently *meaningful*, but it is also in some sense irreducible to a mere mechanical process.

---

19 Rahner, *Hominisation*, 71.
20 This is the idea behind the scholastic Latin phrase *eductio e potentia materiae*, "the leading-out [of form] from the potential of matter."

This aspect of meaning makes Rahner's approach especially useful. In contrast, one could think of generation simply in terms of alteration or unfolding. Geological processes do not really produce new forms of rock; they merely alter the molecular arrangement of atoms. For limestone to become marble is not the creation of something new but an unfolding of an inner potential to exist in a different form. Yet if, on the other hand, there really are developments within nature that arrive at something truly new, then such a generation of newness must have some external, transcendent condition of possibility. If, for example, the transition from proto-ape to what we call "person" is a real, qualitative development—the arrival of something genuinely new—then it must by definition exceed the natural capability of the ape. A lesser does not produce a greater. To use a technological analogy, the pager did not generate the cell phone without the intervention of programmers and technicians.

Nevertheless, theology cannot simply appeal to God to fill this causal gap as though God were merely a cause like any other cause within the world. At a certain time and in a certain place programmers and technicians set about to improve the pager into the cell phone. God is beyond and above time, however, so it makes no sense to limit God's action to "at a certain time and in a certain place." If genuine newness happens in the world, then God must provide this possibility in a way that transcends ordinary causality.

Yet despite being grounded in the causal transcendence of God, the rise of a new form by evolution does not take the shape of a direct, miraculous creation. If an artist reworks a painting by adding something new, then she imparts a fresh, external form to the existing matter. In the case of biological evolution, a new form does not come from an external artist but in some sense arises from within the matter itself. God did not create orohippus by deciding to remake eohippus "at a certain time and in a certain place." Such a divine intervention would be what Rahner calls *categorial*—happening at a specific time and place like any ordinary phenomenon understandable according to Aristotle's categories, as though God were merely one actor among others within the world.[21] Such an approach would be problematic not solely because it relies on God to suspend the ordinary link between cause and effect in the world, but all the more because it restricts the scope of God's creative action, making God seem to be merely one cause among others.

---

21 Rahner, *Hominisation*, 98–99. W. J. O'Hara translates the German *kategorial* as "predicamental."

To be clear, Rahner does *not* claim that God does not intervene in the world. Rather, such direct intervention belongs not to the order of creation but to the order of grace.[22] In contrast, God's creative action must be understood as occurring on a more fundamental level. God is not one cause among others, and therefore even when God creates a new species, there is no seam or gap between ordinary, material causes. God's creative power is neither hindered by nor premised upon the process of biological evolution, for God's causal influence operates on an entirely separate, more fundamental level.

Importantly, Rahner's philosophical account of evolution in terms of generation makes no attempt to explain the particular, scientific phenomena of biological evolution. It is not an explanation of phenomena at all but rather a characterization of the metaphysical realities behind such phenomena. Moreover, it fits far better with a teleological understanding of development than with a true Darwinian dysteleological account of evolution. For these reasons, generation in terms of form and matter appears more patently in the step-by-step evolution of the smartphone from the rotary dial than in the more fluid evolution of *Homo sapiens* from *Homo erectus*. Nevertheless, such an analysis is suited for the task at hand. The concept of spirit is intrinsically a metaphysical reality. It can only be understood from the standpoint of metaphysics, for it is not itself a phenomenon that can be collected, cataloged, and dissected by empirical science.

THE SECOND QUESTION: POLYGENISM AND ORIGINAL SIN

The second issue raised by *Humani generis* concerned the lack of apparent compatibility between the idea of polygenism and the doctrine of original sin. Traditional theology maintains a brand of monogenism, a belief that all humans descend biologically from one original pair, whom the Bible names Adam and Eve. While of relatively little significance today, various forms of polygenism or polyphyletism were strongly in vogue during the early twentieth century. Many scientists, particularly in the field of anthropology, believed that the theory of evolution necessitated polygenism. Growing fossil evidence from earlier forms of humanity seemed unexplainable except by the conclusion that modern humanity evolved several times, in several places, and thus descended from several original pairs.

---

22 Rahner, *Hominisation*, 66.

This issue is greatly exacerbated by the imprecision of the term "polygenism," which long represented a diverse set of views rather than one single hypothesis. *Humani generis* was forced to make use of this term, which was common at the time and frequently served as a direct reproach against the historicity of Adam and Eve. When used in a more precise, technical sense, monogenism and polygenism focus explicitly on the number of original couples (or Adams and Eves). This focus makes the issue inherently theological, since it directly concerns the biblical accounts and the theological implications of our original parentage. This inherently places the issue outside the competence of responsible empirical science, since the bulk of empirical evidence can profess numerical certainty in regard to a quantity of persons that may have existed as many as three hundred thousand years ago.

Within scientific circles, however, the dominant view around the time of the encyclical was really a more refined form of *polyphyletism*. In contrast to polygenism, polyphyletism is a more careful and scientifically responsible concept. It steps back from the numerical debate about individuals and affirms instead the existence of multiple distinct human populations. The evidence in regard to natural selection, after all, is visible principally in regard to populations rather than mere individuals. Polyphyletism has at times implied polygenism, but the two are not strictly bound to one another. Nevertheless, in actual practice, these terms are not always strictly differentiated. In many cases, scholars who insist on polygenism really intend a form of polyphyletism.

Adding to the problem, many forms of polygenism and polyphyletism have in the past been rooted in problematic, nonscientific commitments. It is these nonscientific commitments that make polygenism or polyphyletism potentially untenable from the point of view of Catholic theology. Both positions have a long history of philosophical, theological, cultural, and political implications. They have been influential vehicles of racism, colonialism, and even violence. Understanding this problematic history sheds vital light on the encyclical's second question.

Thus, we turn now to consider the original link between polygenism and racism. Inasmuch as polygenism and polyphyletism are not always strictly distinguishable in this history, we will use "polygenism" here to refer to both except where a more nuanced view is important. This analysis will lead us to appreciate the dramatic difference between earlier, ideological forms of polygenism and more robust and self-aware contemporary ideas about human origins. Separating ideological from authentically scientific commitments will help to highlight what is truly at stake theologically.

Racism and the Origins of the Polygenism versus Monogenism Debate

The roots of the polygenist controversy run deep into the history of early modernity, even before the first stirrings of the Darwinian revolution. It was not empirical science that first gave birth to the suggestion that not all humans are related. Rather, this suggestion arose out of the colonial era as Europeans came to understand themselves in terms of their own supposed cultural, economic, and political superiority. Racism was not a mere incidental byproduct of many early forms of polygenism. It was, in fact, their principal inspiration and aim. To understand this, we must recognize how the concept of race developed alongside polygenism as a justification for European dominance.

Properly understood, both racism and race are authentically modern developments. Of course, ethnic prejudice is as old as human memory. However, the division of humans into distinct, categorizable races is quite recent. Premodern thinking frequently placed people into hereditary and often fluid ethnic categories: Hivites, Amorites, Jebusites, and so forth. All of these could also be reckoned collectively as Canaanites. Yet such groupings were still very local, geographical, and limited; they seldom made absolute claims about the essential nature of a people.

Tertullian (2nd c. CE) provides an interesting example. With his characteristically wry sense of humor, Tertullian has no qualms about mocking Marcion by lambasting the cultural and sexual mores of Marcion's homeland of Pontus: "When they go inside their wagons to have sex, they hang up their quivers on the yoke to warn off anyone who might intrude."[23] Yet Tertullian swiftly moves on to critique the weather: "The daytime is never clear, the sun never cheerful; it's always cloudy, winter year-round." In short, ethnic prejudice operated less in terms of skin color or physiology and more in terms of geography, climate, custom, and morality.

The modern concept of race developed out of the slow coalescence of feudal kingdoms into powerful European nation-states beginning at the end of

---

23 Tertullian, *Adversus Marcionem*, ed. Ernest Evans (Oxford: Clarendon, 1972), 3. My own translation.

the fifteenth century and reaching its apex only in the nineteenth. Not only did this foster among European peoples a broadening sense of identity (e.g., not just as Castilian but now as Spanish), it also led to the growing drive to solidify this identity through colonial dominance. The racial "other" arose as a category that allowed the Europeans to conceptualize their own growing political and cultural hegemonies. They sought to justify the reign of European powers over other peoples by characterizing the "other" as essentially inferior.

Early, less developed racism still relied on traditional ideas for such self-justification. Sixteenth-century Spanish thinkers notoriously argued on the basis of Aristotle that the Native Americans were meant to be slaves by their very nature.[24] Such an Aristotelian approach, however, relied upon neither physiology nor genealogy. It rooted European dominance in the natural order and attacked the intelligence of the natives, but it did not strictly require the Native Americans to represent a different essence or race. Instead, the supposed inferiority of their cultural, religious, and political institutions was sufficient enough to justify their subjugation.

However, as new forms of thought developed across Europe, so also did new ideas of race, which increasingly outlined supposedly essential, absolute, and unchanging differences among peoples as justification for European dominance. By the nineteenth century, for example, many believed in maliciously contrived claims that interracial unions were either infertile or produced monstrous hybrids.[25] This allowed them to think of the difference between white and black in terms of the difference between horse and donkey.

Alongside this development, the modern idea of polygenism first arose in the unassuming work of Isaac La Peyrère (1596–1676). In 1655, La Peyrère published a book arguing on the basis of certain passages in Paul that not all humans were descended from Adam. There were, in fact, "pre-Adamites," who were the forebears of the Gentiles. The lineage of Adam, a privileged subset of humanity, culminates in the Jews. By means of this convoluted genealogical system, La Peyrère sought to justify his own political views.[26] Roundly rejected

---

24 Gustavo Gutiérrez, *Las Casas: In Search of the Poor of Jesus Christ* (Maryknoll, NY: Orbis, 1993), 291–301.

25 J.C. Nott, "The Mulatto a Hybrid—Probable Extermination of the Two Races If the Whites and Blacks Are Allowed to Intermarry," *The Boston Medical and Surgical Journal* 29, no. 2 (August 16, 1843): 29–32.

26 Jeffrey L. Morrow, "French Apocalyptic Messianism: Isaac La Peyrère and Political Biblical Criticism in the Seventeenth Century," *Toronto Journal of Theology* 27, no. 2 (2011): 208–10.

by Calvinist, evangelical, and Catholic authorities alike, La Peyrère's work was of little consequence in itself. Yet it sparked an idea that would continually resurface, particularly in the nineteenth century. A none-too-subtle example can be found in the writing of Alexander Winchell (1824–91), whose voluminous 1880 book serves as a remarkably erudite example of white supremacism. By exaggerating a vast array of stereotypes, depictions, and physiological observations, he argues that Caucasians cannot possibly be close relatives of the other, "inferior" races.[27] In his view, a white Adam represents a higher step of evolution out of primitive, black humanity.

Winchell's opus was the culmination of a complex development of polygenistic thought. Scientific (as opposed to merely theological) polygenism arose in the late eighteenth century. It attempted to justify racial inequality on the basis of what at the time traded for scientific facts. In 1785, the pioneering polygenist Christoph Meiners (1747–1810) insisted that humanity descended from different stocks: the superior Caucasian branch that gave birth to Europeans and the inferior Mongolian branch that originated the other peoples of the world.[28]

Such polygenist positions often found themselves embattled on two sides: on the first, by pious Christians who saw its historical claims as irreconcilable with the book of Genesis; on the other, by more egalitarian theorists who recognized in polygenism a clear denial of the full dignity, humanity, and potential of non-white, non-European humans. Nevertheless, even well-intentioned monogenists did not always succeed in excising racial prejudice from their theories. Johann Friedrich Blumenbach (1752–1840), a colleague and contemporary of Meiners, believed in the basic equality of humanity. Still, he held white, European humanity as the gold standard against which all others were to be measured, and he argued that non-Europeans descended by degeneration or corruption from the original, pure Caucasian stock.[29]

In this way, from the very beginning, the decisive impetus behind the scientific debate over monogenism versus polygenism was racism. Meiners and Blumenbach exhibit the two archetypal lines of racist anthropology: either

---

27 Alexander Winchell, *Preadamites: Or a Demonstration of the Existence of Men Before Adam, Together with a Study of their Condition, Antiquity, Racial Affinities, and Progressive Dispersion over the Earth* (Chicago: S.C. Griggs, 1880).

28 Christoph Meiners, *Grundriss der Geschichte der Menschheit*, 1st ed. (Lemgo, Germany: Meyerschen Buchhandlung, 1785), 25.

29 Wolpoff and Caspari, *Race and Human Evolution*, 62.

humanity is really a collection of originally different populations or species, one superior and the others inferior, or else our unity is founded on a sense of the superiority of one race which is hinted at by the imperfection of all others.

Conversely, the nineteenth century also saw the rise of a more compassionate form of monogenism, as some scholars leaned upon the concept of human unity to combat slavery, colonialism, and deeply entrenched racial inequality.[30] For the most part, Darwin's 1859 publication of the theory of natural selection was received as a confirmation of monogenism; by insisting on the plasticity of organisms, it made it possible to see the anatomical differences among various peoples as having developed from one original stock according to the demands of different environments.

Nevertheless, not every Darwinist agreed. In 1864, Karl Vogt (1817–95) maintained that natural selection made polygenism the only logical conclusion. The distinct races, he argued, evolved separately from different species of apes: "In short, we cannot see why American races of man may not be derived from American apes, Negroes from African apes, or Negritos, perhaps, from Asiatic apes!"[31]

The Downfall of Racist Polygenism

By the mid-twentieth century, the racist roots of the debate often faded into the background. Polygenism, now often refined into polyphyletism, continued to have its staunch adherents. As Vogt's claim that different races evolved from different apes gave way, more robust and evidence-based theories filled the gap. In 1962, Charleton S. Coon (1904–81) argued that modern *Homo sapiens* evolved exactly five separate times from the earlier *Homo erectus*, and that this served as the basis for the division of humanity into separate races.[32] Such theories would continue to fuel racist social and political policies, but many proponents—among them Teilhard de Chardin—were not driven by any racist agenda. Coming from a Christian perspective, Teilhard believed

---

30 Robert Kenny, "From the Curse of Ham to the Curse of Nature: The Influence of Natural Selection on the Debate on Human Unity before the Publication of *The Descent of Man*," *The British Journal for the History of Science* 40, no. 3 (2007): 367–88.

31 Karl Christoph Vogt, *Lectures on Man: His Place in Creation, and in the History of the Earth*, ed. James Hunt (London: Longman, et al., 1864), 466–67.

32 Carleton S. Coon, *The Origin of Races*, 1st ed. (New York: Knopf, 1962).

in the spiritual equality of all humans. His firm attachment to polyphyletism was empirical. He believed that the paleontological evidence demanded polyphyletic reckoning to account for the physiological differences among an ever-increasing array of newly discovered early human and humanlike fossils.[33] In an era before the flowering of genetics, it seemed easier to envision the distinct human populations as having evolved separately and then intermixed to form present-day humanity.

Back in 1950, the future of monogenism had seemed bleak. Teilhard, convinced that the evidence for polyphyletism made traditional monogenism untenable, went so far as to mock *Humani generis* for speaking "with a great deal of high feeling and confusion."[34] Yet Teilhard's adherence to polyphyletism was ill-fated. The decline of the broad acceptance of multi-origin theories came about not for theological but rather for scientific reasons. Teilhard's views, like those of many polyphyletic scientists, were based largely on anthropology and fossil analysis. From this standpoint, Teilhard strongly disagreed with Darwin and natural selection in many ways.

Beginning in the 1940s, however, the rediscovery of genetics gave birth to the neo-Darwinian "modern synthesis."[35] By the 1960s, developments in the science of genetics not only brought important confirmation of natural selection; it also made the likelihood of the human species evolving multiple times "vanishingly unlikely."[36] As Theodosius Dobzhansky demonstrated, even Coon's mature polyphyletism relied on several outrageous assumptions. First, *Homo erectus* would have been unable to breed with the new *Homo sapiens*. The two species lived alongside one another separately for thousands of years, long enough for *Homo sapiens* to evolve from *Homo erectus* four more times. These distinct races of *Homo sapiens* then had to refrain from any interbreeding—despite the human tendency to the contrary—so that each race could be seen as a genetically distinct population even into the modern era. In reality, Dobzhansky argues, five different evolutions would be much more likely to result in five distinct and incompatible species. Yet the genetic differences among humanity's so-called races are actually quite

---

33 Teilhard de Chardin, *Christianity and Evolution*, 46.
34 Teilhard de Chardin, *Christianity and Evolution*, 209.
35 Julian Huxley, *Evolution: The Modern Synthesis* (New York: Harper & Brothers, 1942).
36 Theodosius Dobzhansky, Ashley Montagu, and C.S. Coon, "Two Views of Coon's *Origin of Races* with Comments by Coon and Replies," *Current Anthropology* 4, no. 4 (1963): 365.

small. Such genetic refutation of the once-dominant forms of polyphyletism led to the ascendence today of the monophyletic Out of Africa model—the idea that modern *Homo sapiens* first evolved in Africa and spread to other parts of the world.

Humani Generis and the Theological Side of the Debate

Despite its impending downfall, in 1950 polygenism posed a serious obstacle to the Catholic Church's openness to evolution.[37] The scientific mainstream of the early twentieth century insisted that some polyphyletic view of human origins was necessitated by the evidence, and Teilhard was far from alone in seeing this as running directly counter to the biblical story of Adam and Eve. Yet for the Church, the principal danger of polygenism was *not* that it seemed to contradict the letter of the Bible. Catholic tradition already included a long history of reading much of the book of Genesis as figurative in intention. Rather, the chief problem of polygenism was its propensity for undermining the doctrine of salvation. As we have seen, early polygenistic views, often following racist motives, tended to deny the unity of the human race. This threatened to undermine the universality of the doctrines of sin and salvation. Those who did not descend from Adam either had no share in original sin or else no possibility of sharing in the redemption of Christ.

In view of this, Catholic authorities consistently saw polygenism as mere repetitions of the archetypal heresy of La Peyrère. In 1862, a small, provincial council of bishops from the archdiocese of Cologne decried the opinion "that humans were produced, at least as regards the body, by a spontaneous, continuous, and definitive alteration of a less-perfect nature into this more-perfect humanity."[38] This declaration was penned too soon to have been intended as a condemnation of Darwin, whose *Origin of Species* was published in English only six months prior. Rather, the bishops apparently had in their crosshairs

---

37 Wolpoff and Caspari, *Race and Human Evolution*, 34. For more on theological polygenism, see Augustine Kasujja, *Polygenism and the Theology of Original Sin: Eastern African Contribution to the Solution of the Scientific Problem, The Impact of Polygenism in Modern Theology* (Rome: Urbaniana University Press, 1986), 101–52.

38 *Acta et decreta concilii provinciae Coloniensisin civitate Coloniensi anno domini MDCCCLX pontificatus Pii PP. IX. decimoquarto celebrati* (Cologne: John Peter Bachem, 1862), 30–31 (cap. 9). My translation.

a pre-Adamite view that a supposed lesser form of humanity evolved into a superior (white) form so that only the latter are descended from Adam.

Humani generis effectively continues this same pattern of re-condemning pre-Adamitism. When problematizing polygenism, it explains, "For the faithful cannot embrace that opinion which maintains that either after Adam there existed on this earth true men who did not take their origin through natural generation from him as from the first parent of all, or that Adam represents a certain number of first parents."[39] In this way, the principal target of its problematic is not a scientific perspective but rather a theological one. *Humani generis* does not intend to problematize the specific view that human evolution would have required a larger population than two individuals but rather the view that different human races descend from different original couples, which add up to "a certain number of first parents."

This is not to say that scientific polygenism poses no difficulty, but rather that the encyclical intentionally frames the issue of polygenism in a specific, theological manner, which shifts the discussion away from the particulars of scientific evidence and toward the most vital commitments of Catholic faith. Not every form of polygenism is pre-Adamite, yet pre-Adamitism to some extent illuminates what is ultimately at stake in this issue. Traditional monogenism, by understanding humanity as descended from Adam and Eve without exception, provided a firm theological basis for understanding both how we can all be implicated in original sin and, at the same time, how we can all share in the redemption wrought by Christ. In short, it saw shared biological descent as a concrete basis for the metaphysical unity of the whole human race. Pre-Adamitism and the racist forms of polygenism that followed after it undermined this concept of human unity by separating out the origins of different human races.

Thus, from a Catholic theological standpoint, the real issue within the monogenism versus polygenism debate was always human unity. The question came down to whether or not scientific polygenism (or polyphyletism) dissolved the unity of the human race in a pre-Adamite manner.

Karl Rahner's approach to this issue is illuminating because of its complexity. In 1954, Rahner interpreted *Humani generis* as effectively excluding polygenism and all but mandating monogenism. By 1966, however, Rahner had

---

39 Pius XII, *Humani generis*, §37.

changed his mind.⁴⁰ This was influenced in part by the fact that the Church had not issued any more explicit condemnation of polygenism. Nevertheless, Rahner remained consistent in realizing that at the core this dogmatic issue concerns human unity. There must be a *real* unity among all humans, according to which we can share both in the situation of original sin and in the offer of Christ's redemption.[41] This unity must be a historical reality precisely because original sin and redemption both come to us by way of history.

Rahner details five aspects of human unity. First, there is a basic physical unity inasmuch as the decisions one makes necessarily enter into this one world and impact others in the world. Second, one way or another we derive from the same evolutionary history. Third, we are one species, as demonstrated by our ability to procreate. Fourth, there is a kind of cultural unity, which is not incidental to who we are. Finally, and perhaps most importantly, we are all one in our supernatural destiny toward Christ.[42] These aspects of human unity form the basis upon which it is possible for all humans to share both in one situation of original sin and in one redemption in Jesus Christ. We will further discuss Rahner's approach to these issues in chapter 5.

The Future of Monogenism

The question of monogenism has shifted significantly since the promulgation of *Humani generis*. Theologically, few today have ever heard of pre-Adamitism, and even fewer subscribe to its claims. Scientifically, the debate over human origins has turned away from counting our first parents and toward the question of monophyletism versus polyphyletism. Today's mainstream views are more genuinely rooted in empirical evidence, and as such they really pose much less of a threat to the concept of human unity defended by traditional

---

40 Karl Rahner, "Theological Reflexions on Monogenism," in *Theological Investigations*, trans. Cornelius Ernst, vol. 1 (Baltimore: Helicon, 1961), 249; "Evolution and Original Sin," in *The Evolving World and Theology*, ed. Johann Baptist Metz, trans. Theodore L. Westow (New York: Paulist, 1967), 61–73; "The Sin of Adam," in *Theological Investigations*, trans. David Bourke, vol. 11 (New York: Seabury, 1974), 247–62; "Exkurs: Erbsünde und Monogenismus," in *Theologie der Erbsünde*, by Karl-Heinz Weger (Freiburg: Herder, 1970), 176–223.

41 Rahner, "Monogenism," 277, 279, 284–85; "Sin of Adam," 253.

42 "Evolution and Original Sin," 67; "Exkurs: Erbsünde und Monogenismus"; "Unity of the Church—Unity of Mankind," in *Theological Investigations*, ed. Paul Imhof, trans. Edward Quinn, vol. 20 (New York: Crossroad, 1981), 156.

Catholic monogenism. It has become clear that scientific claims about the evolution of the human race need not deny the fundamental unity that binds us together.

Although the monophyletic Out of Africa hypothesis holds the greatest sway today, it is not the only one in the field. Yet the differences today between respected polyphyletic and monophyletic theories are much more subtle and technical and pose less difficulty for theology.

The multiregional theory of Milford Wolpoff represents a mature, well-evidenced, and robust correction of polyphyletism. Without denying the special significance of Africa, it argues that human evolution cannot be reduced to one simplistic and localized narrative. Rather, over a long history of two million years, distinct populations evolved particular traits in discrete regions across the globe.[43] In this view, all humans (beginning with and including *Homo erectus*) have remained one species, capable of interbreeding, and thus many traits that first developed locally have been shared and spread across the global population. At the same time, some local traits have remained relatively linked to regional populations, so that some distinct characteristics that we associate, for example, with Asians actually trace back to early Asian populations of *Homo erectus*. Based especially on anthropological evidence, this view thus accounts for aspects of *Homo erectus* skeletons that are surprisingly consistent with modern-day humans from the same region.

On the far opposite side of the spectrum is the monophyletic Eve hypothesis. Named for the media's popularization of the phrase "mitochondrial Eve" in reference to it, this theory is rooted in analyses of mitochondrial DNA by Rebecca L. Cann, Mark Stoneking, and Allan C. Wilson.[44] According to the Eve hypothesis, modern *Homo sapiens* as evolved *only* in Africa and as a distinct species genetically *incompatible* with the other hominins, which already inhabited various regions throughout the world. While the Eve hypothesis is a form of the Out of Africa approach, it differs from more moderate forms by insisting that *Homo sapiens* expanded across the globe not by interbreeding but rather by conquest.[45] *Homo sapiens* wiped out other hominins, including

---

43 For an overview of contemporary monophyletic versus polyphyletic theories, see John Hawks and Milford H. Wolpoff, "Sixty Years of Modern Human Origins in the American Anthropological Association," *American Anthropologist* 105, no. 1 (2003): 89–100.

44 Rebecca L. Cann, Mark Stoneking, and Allan C. Wilson, "Mitochondrial DNA and Human Evolution," *Nature* 325, no. 6099 (January 1987): 31–36.

45 Wolpoff and Caspari, *Race and Human Evolution*, 42.

pan-global variations of *Homo erectus*, the fossils of which represent a catalog not of our ancestors but rather of our victims.

From a theological standpoint, the prevailing theories on both sides of this spectrum present at face value less of a theological challenge. In contrast to racist forms of polyphyletism, modern theories maintain (alongside theology) that all humans today are more closely related than they are unrelated. Moreover, even though multiregionalism pushes back the fuller genetic unity of the human race into the time of *Homo erectus*, it is possible from a theological point of view that our first parents were not what science classifies as *Homo sapiens* but rather *Homo erectus*. After all, multiregionalism also minimizes the difference between *Homo sapiens* and *Homo erectus* so that, in terms of reproduction, they still constitute one and the same species. In short, the theological requirement of human unity is not directly contradicted by the terms of today's ongoing debate.

Nevertheless, theological monogenism is not altogether in the clear. Neither the dominance of the Out of Africa hypothesis nor the flexibility of the multiregional theory lends any material support to the specific claim of one historical Adam and one historical Eve. Even the eye-catching references to "mitochondrial Eve" are not intended to defend the Eve of the Bible. From a scientific standpoint, we should not imagine the birth of humanity as beginning with a couple of individuals who suddenly cease to belong to the old species and thereafter populate the new. In reality, evolution typically involves a slow shift in an entire population, as a particular mutation spreads and becomes dominant. The first individual to obtain this new mutation is not a new species. Rather, this person continues to breed with non-mutated individuals. Only a long process of genetic drift would make this mutated population a distinct species, no longer able to breed with its non-mutated forebears. Thus, because evolutionary science concerns populations rather than individuals, it cannot provide support for the historical existence of Adam and Eve.

Moreover, some scientific approaches still directly oppose this idea. Francisco Ayala, for example, provides some of the most challenging evidence against theological monogenism.[46] He points to a gene known as DRB1,

---

46 Francisco J. Ayala, "The Myth of Eve: Molecular Biology and Human Origins," *Science* 270, no. 5244 (December 22, 1995): 1930–36; see Kenneth W. Kemp, "Science, Theology, and Monogenesis," *American Catholic Philosophical Quarterly* 85, no. 2 (2011): 217–36.

which has fifty-nine variants in contemporary human populations. According to Ayala, thirty-two of these variants are also found in analogous (but not identical) form among chimpanzees. If humans and chimpanzees have both preserved these variant forms from a common ancestor, then the population of the species could never have been as low as two. Simply put, two people cannot hold thirty-two variants of the same gene. Mathematically, this would require at least sixteen unique individuals, but realistically the number would have to have been greater, or else some of the variants could have easily died out. Ayala estimates the human population must have at no point been fewer than fifteen to twenty thousand individuals. While Ayala's argument is not without its critics, it does present a significant challenge to the belief that the human population was ever as small as two.

CONCLUSION

In addition to these important theological conclusions, the discussion surrounding *Humani generis* illustrates the central importance of the dialogue between science and theology today. Of course, this remains a complicated task. Theology and science must enter into dialogue precisely because truth is *one*. Since they speak about one and the same truth, each must come to terms with the insights of the other. For Rahner, in particular, this does mean that the Church's Magisterium can in fact make declarations on matters that pertain to science. Yet this is not to say that it does so easily or should do so at all. Rather, it means that genuine dialogue requires a real openness on both sides.

The key point is that science and faith do indeed have their own proper spheres, but in the orthodox Catholic mindset these spheres are not strictly exclusive, as though they can never overlap with one another in any way. This view of course flies in the face of positivism and materialism, and some scientists may understandably fear that it still gives the Church the mandate to interfere with matters that are beyond the understanding of the average clergyman. At the same time, the strength of the Catholic perspective is that it all but necessitates real dialogue between science and theology. Genuine dialogue cannot occur where there is no openness, no possibility of change on either side. Therefore, in the Catholic view, not only must Catholic scientists be open to dialogue with the teaching authority of the Church, but the Church's theology must also be open to being shaped and renewed by the insights of natural science. Evolution, then, can never be something that merely has to do with the fleshy side of the human being. It *must*, from the very beginning, be capable of expressing something deeply true about humanity as a whole, about our world and about the God who created it.[47]

Of course, Christianity holds original sin to be a central element of humanity's identity and the defining status of our relationship with God. If evolution speaks to our very identity as humans, then how might it shed light on this most pivotal of doctrines? We turn now to examine several ways in which evolutionary theologians are rethinking the traditional doctrine of original sin.

---

47 Christoph von Schönborn, "Foreword," in *Creation and Evolution: A Conference with Pope Benedict XVI in Castel Gandolfo*, ed. Stephan Otto Horn and Siegfried Wiedenhofer, trans. Michael J. Miller and Michael J. Miller (San Francisco: Ignatius, 2008), 16.

CHAPTER FIVE

The Evolution of Evil
The Biological Rethinking of Original Sin

As we have seen, the rise of evolutionary theology has produced new and sometimes daring ideas about the meaning and direction of the world, humanity, and even God's own self. Yet we have also seen that much of this evolutionary thinking really relies upon earlier questions and insights, which thinkers take up anew in light of a deepening evolutionary framework. It was neither Whitehead nor Darwin who first lighted on the problem of evil; likewise, it was not evolution that first made the doctrine of original sin controversial. Nevertheless, original sin quickly became one of the key areas of debate, and a site upon which often hinged human evolution's very acceptability among Christians. Since its inception, evolutionary theology has developed (or revamped) not one but many reinterpretations of the doctrine in light of evolutionary theory. In fact, as Raymund Schwager argues, the evolutionary rethinking of original sin—despite the frequency of problematic views—turns out to be vital for conceptualizing the doctrine today precisely because it reveals what is most at stake. Whether or not one finds such contemporary reinterpretations dogmatically tenable, facing the critiques and consequences of evolutionary theology is key to retrieving the doctrine of original sin today.

In order to illuminate this significance, we must examine the real historical roots of the controversies surrounding original sin. A concise but careful look illustrates how the sixteenth-century Reformation's reconceptualization of the doctrine of original sin established the key separation between freedom and nature, which will become the basis for many modern Protestant ideas about original sin. This separation, alongside developments in science and philosophy, will spur the formation of evolutionary rereadings of the doctrine. We will explore three such reinterpretations in particular, examining their key contributions and offering a few critiques. Finally, we will look at how Karl Rahner and Raymund Schwager refocus on the historicity of original sin in order to provide a more tradition-centric reinterpretation of the doctrine.

THE HISTORICAL ORIGINS OF THE DEBATE

Original sin has always been controversial. Because of this, contemporary critics have no difficulty questioning its historical roots. Some go so far as to uncritically accuse Augustine of Hippo (354–430) of having concocted the doctrine out of nothing. Yet while Augustine did indeed provide its first complete, systematic formulation, he saw this doctrine not as something novel but rather as an authentic reading of the teachings of Paul, the book of Genesis, and prior theological tradition. The relatively swift acceptance of Augustine's formulation shows that Augustine was far from alone in his belief that humanity was responsible for the advent of sin and that this responsibility constituted an ongoing burden, which could only be expunged by the death and resurrection of Christ. He fit neatly into a long Jewish and Christian tradition of rereading the book of Genesis in light of shifting theological perspectives (see Wis 2:23–24).

For Augustine, the doctrine of original sin essentially communicated a message of hope. It defied the Manichees' belief that some people were irreparably evil, and it confounded those hardcore ascetics who saw spiritual success or failure as a matter of individual character alone. The doctrine brought hope to the average, unexceptional Christian. Anyone can attain to the spiritual heights precisely because, in the final analysis, one's triumph over sin does not hinge upon one's own skill, refinement, intelligence, or willpower but rather upon the undeserved gift of divine grace. To say that everyone is implicated in the history of sin is to say that no one is a lost cause. This runs directly counter to our common human tendency to elevate oneself at the expense of others, to hold oneself to be fundamentally innocent by casting another as guilty. Salvation is not a matter of spiritual genius or superhuman effort. It is rather the drama—the romance—of submitting oneself to the merciful assistance of a loving God.

We cannot treat in this short space the full breadth of Augustine's teaching along with its proper historical context.[1] Let it suffice to indicate that while the broad adoption of this doctrine within the Church shows that while it was far from alien or the invention of a single man, neither was it entirely uncontested. As in other major controversies—most notably that of Arius—contradictory threads within the complex tapestry of Christian tradition found themselves at odds. Augustine's principal opponent, Pelagius, formed his arguments out of Christian tradition just as much as Augustine. Pelagius even quoted Augustine's mentor Ambrose of Milan in support of his own views.[2] The basis of Pelagius's argument was the time-honored ascetic tradition, which emphasized personal purification through self-denial, self-isolation, and even flagellation. If Christians have long practiced celibacy, fasting, sleep deprivation, and other techniques of taming the body for the sake of the spirit, then it must mean that personal effort was really capable of achieving salvation through hard work. Augustine's doctrine of original sin seemed to Pelagius to threaten such practices and even to encourage spiritual laxity by removing the work of salvation from the realm of free-will and making it exclusively a matter of divine gratuity.

Yet where Pelagius saw a dichotomy, Augustine perceived a deeper unity. The necessity of grace does not undermine the significance of human freedom, he argued, but rather confirms it. Because we are implicated in Adam's sin, we inherit a basic misguided tendency toward sin: *concupiscence*. In light of this, we are not born into our natural freedom, but find ourselves the would-be slaves of sin. It is grace that actually enables us to be free by liberating the free-will and moving it toward the ultimate good that is God. Human free-will is still valuable and necessary for salvation, but in no way does it achieve salvation of itself without the prior help of grace. Ascetic practices are beneficial not because they accomplish spiritual growth through sheer will

---

1   Some detailed analyses of Augustine's teachings include: Charles Baumgartner, *Le péché originel* (Paris: Desclée de Brouwer, 1969); Jesse Couenhoven, "St. Augustine's Doctrine of Original Sin," *Augustinian Studies* 36, no. 2 (2005): 359–96; Paul Rigby, *Original Sin in Augustine's Confessions* (Ottawa: University of Ottawa Press, 1987); Henri Rondet, *Original Sin: The Patristic and Theological Background*, trans. Cajetan Finegan (Staten Island, NY: Alba House, 1972).

2   Augustine, *On the Grace of Christ and Original Sin*, book I, ch. 43, 47. J. Patout Burns highlights the elements of tradition underlying Pelagius's views in his anthology: *Theological Anthropology* (Philadelphia: Fortress Press, 1981).

and determination but rather because they embody the actions of a free-will already lovingly liberated by divine grace.

Of course, not everyone was convinced. Despite Augustine's understanding of the doctrine achieving relative dominance especially in the West,[3] contrary theological threads persisted into the Middle Ages. It was not until the Protestant Reformation, however, that these threads really came to the fore and caused a new plurality of ways of conceptualizing the doctrine. Ultimately, the Reformation's rehashing of the problem of original sin will lead to the very questions that shape the intersection between the doctrine of original sin and the idea of human evolution.

The Reformation Shifts the Categories for Conceptualizing Original Sin

While the Reformation began in stages as early as the fourteenth century with the movements of John Wycliffe (1320–1384) and Jan Hus (1372–1415), the oft-dramatized narrative cherishes the image of a defiant Martin Luther (1483–1546) nailing his *Ninety-five Theses* to the door of a university chapel in 1517. As melodramatic as this representation is, it highlights the unique significance of Luther, whose views came to exert a broad but varied influence on the many nascent forms of Protestant Christianity. It is in Luther's theology that the controversies around the doctrine of original sin again came to the fore, and it is his arguments on the issue that will shape the landscape of this debate into the modern era and up to the present day. Not that Luther

---

3 The now commonplace assertion that Eastern Christianity never accepted the doctrine of original sin is fueled more by bias than evidence. While it is true that a fully Augustinian concept of original sin—involving inherited guilt, concupiscence, and the absolute need for grace—is far from universal among the Eastern fathers, key aspects of Augustine's doctrine can be seen for example in the teaching of Gregory of Nazianzus, whom Augustine even cites as an authority. That this doctrine gained purchase among many Eastern Christians by the modern era can be seen in the teachings of the renowned Patriarch of Moscow, Saint Filaret (1783–1867): *Select Sermons* (London: J. Masters, 1873), 138, 145, 175–76, 311–15, 360; "The Longer Catechism of the Russian Church," in *The Doctrine of the Russian Church* (London: J. Masters, 1845), 60. At least two recent Orthodox catechisms present views even more explicitly in line with Catholic teaching: George Mastrantonis, *A New-Style Catechism on the Eastern Orthodox Faith for Adults*, 2nd ed. (St. Louis, MO: The Ologos Mission, 1977), 73–80, 93–94; and Carl S. Tyneh, ed., *Orthodox Christianity: Overview and Bibliography* (New York: Nova Science Publishers, 2003), 47–53.

was altogether innovative in this regard; like Augustine and Pelagius before him, he gave important expression to theological trends that were already present within the milieu of Christian thought.[4] Most importantly, he shifted the focus of the doctrine toward a staunch insistence on specifically *personal* responsibility, sowing the seeds of a conflict between nature and freedom that will flower with the rise of evolutionary thinking.

By the sixteenth century, the authoritative Catholic interpretation read Augustine's doctrine more along the lines of communal than of personal guilt. The original sin that we inherit (*peccatum originale originatum*, originated original sin) is not an *actual sin*—that is, it is not a sin personally and willfully committed by an individual, for which that same individual is duly responsible. Rather, we share only *analogously* in the guilt of our first parents, whose own actual sin (*peccatum originale originans*, originating original sin) gave birth to this history of guilt. Something like guilt is inherited, but not personal guilt. We are not guilty for original sin in the same way that Adam was. While I am born guilty, I have not personally wronged God, nor does this juridical guilt (Latin *reātus*) require that I should feel personally guilty (*cōnscīre*). Rather, the concept of inherited guilt serves to underline my basic human need for God's saving grace. It means, on the one hand, that God is not responsible if I receive damnation—I have always deserved it from the point of my conception—and on the other, that God alone is the true author of my salvation, even if God generously allows me to share in the merits of my good deeds.

Concurrently, through the influence of Anselm of Canterbury (1033/34–1109) and Thomas Aquinas (1225–74), Catholic teaching understood our inherited situation of *peccatum originale originatum* less as a positive accretion (like a stain on a T-shirt) and more as a negative deprivation of something that otherwise should be (the loss of the shirt's rightful cleanliness). God had provided humanity with grace immediately upon creation—the *grace of original justice*. When our first parents sinned and lost this grace, its catastrophic consequences thrust us into a situation far worse than if we had never received grace at all. In this sense, original sin represents the loss of the grace of original justice. It is not some artificial blame levied by God against humanity but rather the organic consequence of how sin severs the

---

4 In particular, elements of Luther's emphasis can be seen in Anselm, *On the Virgin Birth and Original Sin*.

relationship between God and humanity. Grace, after all, is God's intimacy with us, and to lose grace is to cut ourselves off from that relationship.

These views on original sin remain dominant within Catholicism to this day. Importantly, from this standpoint, our present-day situation of original sin (*peccatum originale originatum*) is the natural consequence of the historical deed (*peccatum originale originans*) of our first parents. The personal deed of real humans has had a lasting impact on history, irreparably shaping the situation in which each and every human enters into this world. It is a burden that impacts each of us on a personal level, a situation with which I must inevitably contend, but not thereby a matter of personal, individual responsibility.

For Martin Luther, however, this understanding of the doctrine did not do enough to counter the hubris of human effort. In his drive to combat the Catholic theology of *merit*—the idea that we can share credit for the good deeds that we accomplish by the grace of God—Luther sought to rehash original sin in a way that would drive home an absolute dichotomy between God and humanity, grace and free-will, faith and reason. Whereas Catholicism saw human nature as wounded and weakened by the fall, Luther saw it as irreparably evil, destroyed, and vitiated.[5] As a result, an individual is utterly incapable of accomplishing any good whatsoever apart from grace.

Most importantly, Luther insisted that if we are guilty of original sin, then we must each be *personally* guilty. In other words, original sin is *actual sin*. It is not just something that Adam did but also something that I myself have done. Luther saw confirmation of this in our inborn inclination toward sin (concupiscence). He reasons that one's mere possession of such an inclination must itself be sinful, and thus I must be personally responsible for it. Since I am utterly destitute and incapable of any good, the only solution is for me as an individual to accept God's grace by means of faith.

Notwithstanding a variety of particular forms, Luther's emphasis on personal responsibility achieved a certain dominance among the Protestant traditions. It helped to show how original sin specifically impacts the individual believer. It also sidestepped the difficulty of explaining how the historical guilt of our forebears could be transferred to their descendants. Nevertheless, it raised a new and equally difficult question: when and how did I personally sin in order to become personally guilty of original sin? For example, did I sin

---

5 Martin Luther, *Disputation against Scholastic Theology*, 1:9.

as an infant in the womb? If original sin is my own personal choice, then is it possible that someone could be born, live, and die without ever having sinned at all? In other words, how can the inheritance of original sin be unavoidable if I must personally sin in order to inherit it?

Luther was hardly concerned with giving a concrete and complete answer to these questions. In his mind, the ultimate proof of this state of sin lay in the very biblical doctrine of salvation. If the Bible tells us that grace is necessary, then I must be personally guilty; the very offer of mercy confirms my initial state of damnation. The truth about original sin must ultimately be accepted on faith. Faith alone illuminates the futility of our attempted good works.

Still, at times Luther also filled the gap with other arguments, which would be taken up variously by later Protestant theologians.[6] First, he suggested that sin is unavoidable because each and every action that we attempt, if not produced by grace, is a sin. Since only God can accomplish good, even our good intentions produce only evil on their own. In this way, even one's very first breath could be seen as a sin, since it is not a product of grace but rather the fruit of a destitute and rotten human nature. Alternatively, Luther also suggested that our actions remain sinful—even after the reception of grace—to the extent to which they remain imperfect. When I attempt a good deed, I cannot altogether resist ulterior motives such as looking good in order to obtain the favor of others. Consequently, even while the goodness of the deed is credited to the grace of God, it stands on my part as a sin, a failure to do good rooted in my own selfish motives. The doctrine of original sin thus meant for Luther that one way or another I am personally guilty. Either God's mercy proves that I am already sinful without having done anything in particular, or everything I do is a sin, or else I sin by failing to act out of pure and unadulterated motives.

Luther's intention is altogether clear. By establishing the wretchedness of the human race, he set up a firm ground for insisting on the need for divine grace obtained purely by the act of faith and not at all as the product of personal good works. This teaching carries a homiletic force that drives home Luther's call for conversion. Such language goes hand in hand preaching the wrath of an angry God—a common strategy among many of Luther's intellectual descendants.

---

6 Martin Luther, *Disputation against Scholastic Theology*, 5:62–65, 76–78. See also his *Heidelberg Disputation*. Both can be found in *Early Theological Works*, trans. James Atkinson (Philadelphia: Westminster, 1962).

Jean Calvin (1509–1564) presents the significance of this message in straightforward terms. While it is true that God loves us and approaches us with mercy, for Calvin it is more effective to preach first and foremost a message of divine hatred so that an individual might realize one's own guilt and immediately seek divine help.[7] In short, the function of the doctrine of original sin has decisively shifted into a rhetorical realm intended more to impel us to conversion than to rationally explicate the state of fallen humanity.

The Protestant turn toward actual guilt fits into the overarching historical trend toward subjectivity that matured into the Enlightenment. Modernity, as the offspring of the Enlightenment, is dramatically shaped by this "turn to the subject." As such, modern thinkers tend to continue Luther's line of inquiry, treating original sin either primarily or exclusively as a personal, subjective situation of the individual *as* individual rather than any kind of communal inheritance.

The Modern Dilemma: Freedom vs. Nature

In the modern era, the conceptual problems surrounding the doctrine of original sin intensified along the very lines set by Luther's position. Modern thinkers tended to take the distinction between freedom and nature as axiomatic. From the traditional Catholic standpoint, the personal, free act of Adam and Eve has shaped our very nature for the worse. Moreover, because we inherit concupiscence, our own free decisions are also shaped by our fallen nature. For modern critics, however, such claims blur the distinction between freedom and nature by making each open to the influence of the other. "Absolutely contradictory," writes Adolf von Harnack in 1890, "are the positions that all sin springs from freedom (the will), and that children just born are in a state of sin."[8] Many Protestant thinkers sought to rearticulate the doctrine of original sin in a way that would safeguard this fundamental distinction. The origin of human evil must either fall entirely on the side of freedom or else entirely on the side of nature.

---

7 Jean Calvin, *Institutes of the Christian Religion*, ed. John T. McNeill (Philadelphia: Westminster, 1960), book II, ch. 16, sect. 2.

8 Adolf von Harnack, *History of Dogma* (London: Williams and Norgate, 1894), 5:219–20.

Locating the origin of evil within nature sufficiently explains the universality of sin, but it intensifies the difficulty of explaining how one can reconcile the existence of evil with the goodness of God. If nature necessarily contains evil, and if God is the author of nature, then is not God the first and original sinner? Gottfried Wilhelm Leibniz addressed this problem by limiting God's power. God could not have created the world otherwise. Since God necessarily does what is best, this world must be the very best that God could create. God cannot make an omelet without breaking a few eggs; the inevitable existence of evil must in some way be necessary for the existence of good. In fact, it must have been better for God to create a world with evil than to do nothing at all. "To permit evil, as God permits it," writes Leibniz, "is the greatest good."[9] Hegel goes much further. "For the very notion of spirit is enough to show that man is evil by nature," he writes, "and it is an error to imagine it otherwise."[10] Yet Hegel places evil on the side of nature precisely so that the dialectical interrelation of nature and freedom might redeem it. Inasmuch as nature is a necessary component in the coming-to-be of Spirit, so also evil is a necessary component in the coming-to-be of the good.

Such nature-oriented approaches to original sin received new vigor with the publication of *The Origin of Species*. The discovery of natural selection by Wallace and Darwin led credence to the idea that evil is necessary to nature's very existence. After all, violence, suffering, struggle, and even death are vital to the process by which more-fit populations dominate and the less-fit fade away. Moreover, by explicitly linking the concept of original sin to the natural world, such explanations carried the added benefit of explaining the existence of natural evils alongside moral evil. Animals kill each other because an evolving world is inherently imperfect. Likewise, humans kill one another because we have only imperfectly transcended our base animal instincts toward competition and violence. As we shall see, the evolutionary approach to original sin will reach a new apex when later theologians, inspired by the idea of the "selfish gene," project the origin of evil further back into the evolutionary process itself.

Such nature-focused views on original sin have no difficulty explaining sin's universality. Nevertheless, they struggle to explain how this universal situation of evil is actually sinful. If one has no choice in the matter, then

---

9 Leibniz, *Theodicy*, 88.
10 Georg Wilhelm Friedrich Hegel, *The Logic of Hegel*, trans. William Wallace, 2nd ed. (Oxford: Clarendon Press, 1892), 56.

clearly one cannot be held responsible for it. Instead, Leibniz and others tend to hold personal sin as an entirely different matter. Original sin is a metaphor for the situation of evil inherent in the very nature of the world. It has no bearing on the actual culpability of a human person except inasmuch as it may provide opportunities or provocations for us to misbehave. Original sin is original, but it is not really sin.[11]

On the other side of the spectrum, thinkers who located original sin entirely on the side of freedom faced a different conundrum. If original sin is entirely one's own personal choice, then how can it be universal? Is it not possible that someone could avoid sinning altogether? This posed a problem not only because Christian tradition held original sin to be universal, but also because this dogmatic universality served to underscore the moral imperative of the doctrine.

Immanuel Kant attempted to solve this conundrum by rooting the necessary existence of evil not within nature but rather within the very faculty of freedom. Each person's freedom carries an intrinsic propensity for either good or evil, yet this inborn propensity is the product of the individual's own free decision.[12] There is a kind of fundamental, transcendental decision that we make before all other decisions. Eventually, all of our other decisions can be traced back to this transcendental decision, but this decision has no cause other than freedom itself. Ordinary, concrete decisions can always be explained by prior motives. I may choose to steal a pie because I desire its goodness. This transcendental decision, on the other hand, has no cause other than free decision itself. Our capability of evil is thus rooted in something without a root. Kant terms this rootless root *radical evil*. ("Radical" comes from the Latin *radix*, "root," and is related to the word "radish.") In some sense, Kant sees radical evil as a necessary concomitant of human freedom; to be free is to have radical evil. This means that evil is not a part of human nature, properly speaking, but rather an unavoidable byproduct of human freedom.

---

11 See for example Charles Birch and John B. Cobb Jr., *The Liberation of Life: From the Cell to the Community* (Cambridge: Cambridge University Press, 1981), 120.

12 Immanuel Kant, "Religion within the Boundaries of Mere Reason," in *Religion and Rational Theology*, ed. Allen W. Wood, trans. George Di Giovanni (Cambridge: Cambridge University Press, 1996), 76–78. For a good summary, see Robert F. Brown, "The Transcendental Fall in Kant and Schelling," *Idealistic Studies* 14 (1984): 49–66.

Kant obviously stops short of any further explanation because if radical evil were to have any prior cause, then it would become God's fault. If, for example, our transcendental decision for evil were due to some defect in human freedom, then God would be responsible as the author of that freedom.

F. W. J. Schelling illustrates the danger of taking Kant's approach to the extreme. Unsatisfied with Kant's refusal to push further, Schelling traces the origin of radical evil within God's very self. He agrees that radical evil comes from human freedom. Yet he adds that even before human freedom came to exist, radical evil preexisted as an unavoidable element of God's own free decision to exist. Before all of creation, God's fundamental decision (*Entscheidung*) to come into being and create the world was premised upon God's own inner nature or "dark ground," a riotous swelling and swirling of contradictory forces that is just as fundamental as God's freedom. This "dark ground" is something "in God which is not *God himself*."[13] It is a fundamental and unavoidable chaos upon which God must premise not only the existence of the world, but even the very existence of God's own self. Such chaos is not evil per se, but it does mean that contradiction stands at the very basis of reality. Being can only exist because of non-being *and vice versa*; freedom can only exist because of nature *and vice versa*; so also good can only exist because of evil *and vice versa*.

Both modern approaches to original sin—through nature and through freedom—continue to exert significant influence on theology today not only among Protestants but also among many contemporary Catholic theologians. And although these modern approaches differ fundamentally, they come together in their essential conviction that original sin must fall either on one side or on the other. In other words, they take for granted the axiomatic separation and mutual inviolability of nature and freedom.

This brief overview of the various trajectories of original sin in the modern era has put us into a better position for exploring in greater depth some of the key views of contemporary evolutionary theology. From here, we can see how the question of theodicy, especially intensified under the influence of process theology, has led some to see evil as rooted in nature in even more radical ways. After this, we will look back at Catholic theology for an important rejoinder. Catholicism's emphasis on historicity connects with evolution while rejecting the stark modern disjunction between nature and freedom.

---

13 Schelling, *Of Human Freedom*, 33–34.

THREE EVOLUTIONARY NARRATIVES OF ORIGINAL SIN

As we have seen, the mainstream of Protestant theology framed the doctrine of original sin in terms of either the corruption of impersonal nature or the rebellion of personal freedom. With the rise of modernity, the expanding array of modern views tended to fall on a spectrum. One extreme exonerated the individual's conscience by locating original sin within the raw, irresistible necessities of nature. The other emphasized individual responsibility by seeing original sin as something utterly personal, willful, and uninherited despite its universality.

Both extremes were equally unconcerned with the sin of Adam, and thus both saw an ally in the rise of evolutionary science, which tended to undermine belief in the actual, factual existence of our first parents. Nevertheless, nature-focused versions of original sin benefited more from evolutionism. The more a theologian emphasizes individual freedom as the site of original sin, the less the doctrine has anything to do with history. Thus, freedom-focused approaches to original sin have little to no interest in mapping sin onto an evolutionary process. They reason that for as long as humans have been humans, sin has been sin, and it will continue to be the same for as long as freedom remains freedom. In contrast, if we locate the origin of sin within the realm of nature, and if by way of evolutionism we see nature itself as a process of development, then this raises again the possibility that sin has a history. If sin in some sense belongs to our natural past, might its eradication be the work of the future? Might we see the ongoing process of evolution as a movement away from sin and toward greater freedom?

The idea of evolution thus hints in directions that reconceptualize original sin in terms of a dynamic historical trajectory. In order to illustrate the possibilities, we will look in particular at three common evolutionary narratives of original sin. These are conceptually distinct, but little prevents a thinker from adopting multiple at the same time. First, there is the view of sin as a kind of cosmic immaturity. In relation to this, history is a gradual ascent away from the immaturity of sin and toward a higher good. A second, more extreme version of this same narrative is the Gnostic inversion of the fall, which sees sin as "falling up" toward a higher good. Last, there is the growing field of "selfish gene" theology, which sees evil not as part of objective nature but rather as a prior, formal principle governing the evolutionary process itself. After exploring these three narratives, we will consider an important critique from the angle of Catholic tradition.

History as Gradual Ascent

The first narrative sees history as a gradual, upward development toward a higher state of existence. In an 1886 example, the Unitarian orator Minot J. Savage (1841–1918) proclaims that in order to accept evolution "you will have to surrender your belief in 'the fall of man.' Evolution teaches *the ascent of man*; that the perfect Adam is ahead of us, not behind."[14] Evil and suffering, he argues, are really integral aspects of God's plan for training the fledgling human character. Savage thus also rejects the existence of hell, insisting that God punishes no one. Rather, we suffer the consequences of our own bad character.

Today, this upward ascent narrative finds its most influential retelling in the works of the Presbyterian theologian John Hick. His approach is deeply shaped by modern philosophical and theological sources, but Hick prefers to attribute it to Irenaeus of Lyon (ca. 120–202 CE), whom he sees as representative of a separate tradition within Christianity that runs counter to the dominant Augustinian emphasis on the fall.[15] According to Hick, Irenaeus understands human sin to be the result of Adam and Eve's initial immaturity. In order for them to achieve true, personal maturity, God had to create them in a childlike state in which sin would be inevitable. Yet this immature condition also provides the possibility of authentic spiritual growth or "soul-making." History is thus the gradual process in which the human race comes to full spiritual and moral maturity through the struggle with sin and evil. God freely allows evil in the world so that we might accomplish the higher good of spiritual development.

Hick follows Milton, Leibniz, and others by assuming that the Augustinian doctrine of original sin is first and foremost a theodicy.[16] Rather than seeing Augustine's teaching as a message of hope, he caricatures it as a "justification of the ways of God to man." Not unsurprisingly, Hick finds this theodicy

---

14 Minot J. Savage, *Evolution and Religion, from the Standpoint of One Who Believes in Both* (Philadelphia: G.H. Buchanan, 1886), 43; see also *Religion for To-day* (Boston: G.E. Ellis, 1897), 119–35; and *The Religion of Evolution* (Boston: Lockwood, Brooks and Co., 1876).

15 John Hick, *Evil and the God of Love*, 1st ed. (New York: Harper and Row, 1966), 217–21.

16 On Augustine's theodicy, see Johannes Brachtendorf, "The Goodness of Creation and the Reality of Evil: Suffering as a Problem in Augustine's Theodicy," *Augustinian Studies* 31, no. 1 (2000): 79–92.

lacking, since for example it does not really account for the existence of nonmoral evils or the death of animals.

As for Hick's appeal to Irenaeus, this is more strategic than material. Irenaeus certainly does make a few remarks about humanity's immaturity,[17] but the majority of his treatment of the fall squares surprisingly well with Augustine. In reality, the primary source behind Hick's soul-making narrative is not Irenaeus but Friedrich Schleiermacher (1768–1834).[18] This explains why Hick finds it easy to deny the existence of hell[19]—a denial Irenaeus would certainly have scoffed at. In this way, despite lacking significant reference to evolution, Hick's narrative is through and through modern and a product of many of the same influences behind evolutionary theology.

In addition to its widespread appeal among many evolutionary theologians, Hick's approach is noteworthy because he is critical of process theology and, in conjunction with this, intentionally stops short of our second narrative, the Gnostic inverted fall. Recall that process theology understands God to be inherently self-revealing precisely because it is through such self-revelation that God becomes God. Process theologians thus typically account for evil in the world by denying God's omnipotence; like a child, God is in a process of growth and must suffer the growing pains of the world. In contrast, Hick intends in his theodicy to retain an element of mystery, so his approach is not meant to provide a comprehensive and fully logical solution to the problem of evil.[20] It provides perhaps an understanding more than a justification. Connected with this, Hick maintains God's omnipotence but sees God as having voluntarily set it aside in order to cultivate the world's freedom. In order for humans to come to God freely, he reasons, God's very existence must be veiled. God must create a world that seems "as if there were no God."[21]

---

17 Irenaeus, *Adversus Haereses* IV, 38, 1.
18 Hick, *Evil*, 225–41; Mark S.M. Scott, "Suffering and Soul-Making: Rethinking John Hick's Theodicy," *Journal of Religion* 90, no. 3 (2010): 313–34. Cf. Michael Reeves and Hans Madueme, "Threads in a Seamless Garment: Original Sin in Systematic Theology," in *Adam, the Fall, and Original Sin: Theological, Biblical, and Scientific Perspectives*, ed. Hans Madueme and Michael Reeves (Grand Rapids, MI: Baker Academic, 2014), 213–14.
19 Hick, *Evil*, 263; see also *Death and Eternal Life* (New York: Harper and Row, 1976), 198–201.
20 Hick, *Evil*, 371–72.
21 Hick, *Evil*, 317. See also C. Robert Mesle, "Does God Hide from Us?: John Hick and Process Theology on Faith, Freedom and Theodicy," *International Journal for Philosophy of Religion* 24, no. 1/2 (1988): 93–111.

More recently, Niels Henrik Gregersen (1956–) has promoted an influential Lutheran version of the upward ascent narrative. Traditional interpretations of Luther typically push back against such a narrative both because of Luther's stark, negative attitude toward the world and because his strong emphasis on the necessity of grace does not typically fit well with an optimistic view of history. Nevertheless, Gregersen follows Reformed theologian Jürgen Moltmann (1926–) in subjecting Luther to significant critique and modification.

For example, Moltmann utilizes Luther's insistence that Christ was actually forsaken by the Father on the cross to argue that in the crucifixion event, God actually suffers in God's own self.[22] The divine Son suffers a God-forsakenness that goes down to the depths of his divine existence. For Moltmann, this means that God suffers alongside God's creatures so that, without diminishing the absurdity and injustice of suffering, this co-suffering provides a comforting answer to the problem of theodicy. In fact, Moltmann sees Christ as bringing redemption not only to humanity but even to the natural world and "the victims of evolution," i.e., those creatures whose suffering and death was necessitated by the process of evolution itself.[23]

Gregersen similarly argues that Christ's suffering on the cross is symbolically relevant for the natural world. Gregersen insists that the sufferings due to natural processes (natural selection, disasters, etc.) are not a consequence of human sin but rather natural concomitants of God's creative process.[24] Once such a distinction is made, natural suffering and painful competition can be seen as "the price to be paid" for the positive benefits of evolution. God does not turn a blind eye to such suffering. Rather, Gregersen argues, the death of Christ on the cross has salvific significance not only as a response to human sin but also for natural evils and social injustice. "The death of Christ," writes Gregersen, "becomes an icon of God's

---

22 Jürgen Moltmann, "The 'Crucified God' and the Trinity Today," in *New Questions on God*, trans. David Smith (New York: Herder and Herder, 1972), 34–35. See for example Luther, *Early Theological Works*, 50. Calvin develops this emphasis: Calvin, *Institutes*, II, chap. 16, §11. On the related topic of the death of God in Martin Luther, see Frederiek Depoortere, "'God Himself Is Dead!' Luther, Hegel, and the Death of God," *Philosophy and Theology* 19, no. 1/2 (July 1, 2007): 171–195; Dennis Ngien, "Chalcedonian Christology and Beyond: Luther's Understanding of the *Communicatio Idiomatum*," *The Heythrop Journal* 45, no. 1 (2004): 54–68.

23 Jürgen Moltmann, *The Way of Jesus Christ: Christology in Messianic Dimensions*, 1st HarperCollins ed. (San Francisco: HarperSanFrancisco, 1990), 294–97.

24 Niels Henrik Gregersen, "The Cross of Christ in an Evolutionary World," *Dialog* 40, no. 3 (2001): 196–97.

redemptive co-suffering with all sentient life as well as with the victims of social competition."²⁵

Like Hick, Gregersen does not intend to provide a complete answer to the problem of evil. Moreover, while Gregersen does ascribe a real suffering to the divine on the cross, the efficacy of this suffering is merely symbolic. Jesus is the "*icon of a loser* in the evolutionary arms race" and the "*icon of an outlaw*" who refused to play society's unjust game of competition.²⁶ In other words, Gregersen's approach stops short of any Hegelian claim that God requires suffering in order to become more fully God.

The Gnostic Inverted Fall

The second narrative, a more radical version of the upward ascent, is the inverted fall. In its mildest form, this narrative sees the development from proto-ape to human as a kind of fall into something higher.²⁷ This frequently leads, however, to a claim that sin, in the broad scheme of things, is actually necessary and perhaps even good.

By comparison, Catholic tradition is able to refer to original sin as "O happy fault!"²⁸ only by way of irony: the unfortunate and even horrific circumstance of sin, ironically, led to the gracious incarnation of the Son of God. The sin of Adam is accordingly necessary not in any absolute sense but only as the material occasion against which God happened to bring about a categorically greater good. The inverted fall narrative, in contrast, effaces the irony of such a statement. It argues that even if sin remains problematic, it is nevertheless necessary in a more fundamental way.

The Gnostic heresies (ca. second century CE) are archetypal here. They saw the very creation of the world as an evil but necessary fall away from the original self-enclosed divinity of the Godhead.²⁹ God did not desire that the

---

25 Gregersen, 205. Cf. George L. Murphy, "Cosmology, Evolution, and Biotechnology," in *Bridging Science and Religion*, ed. Ted Peters and Gaymon Bennett (Minneapolis, MN: Fortress Press, 2003), 210–12.

26 Gregersen, "The Cross of Christ in an Evolutionary World," 203–204.

27 Gabriel Daly, *Creation and Redemption* (Wilmington, DE: M. Glazier, 1989), 139–41; Joseph Fitzpatrick, *The Fall and the Ascent of Man: How Genesis Supports Darwin* (Lanham, MD: University Press of America, 2012), 200.

28 A famous line from the Easter proclamation known as the *Exsultet*.

29 The most detailed record we have of Gnosticism comes from Irenaeus's condemnation of it in *Adversus haereses*. As far as secondary sources go, the work of Hans

world should exist, but a kind of imbalance within God led to a part of God falling away and becoming entrapped within the material world. In view of this, the Gnostics saw the Creator God of the Old Testament as an evil pretender, a false or would-be God. If the Old Testament God is evil, then of course eating from the tree of knowledge was really the archetypal act of righteous civil disobedience. As a result, Cyril O'Regan explains, "the serpent in Gnostic texts is, at a minimum, both hero and truth teller, at a maximum, the savior as the figure of gnosis."[30] O'Regan's thesis is that Hegel retrieves and revives Gnosticism as a key to reconfiguring the Christian tradition. In fact, Hegel's own inversion of the fall illustrates the importance of this narrative. Inasmuch as human history is necessary for the process of divine Spirit, the fall for Hegel is really a necessary and pivotal event in the history of God's own self. In this way, inverting the fall makes sin necessary and beneficial for humans and God alike.

The pattern of the Gnostic inverted fall entered into evolutionary theology quite easily. If humans evolved from a proto-ape, then ironically, there is a sense in which to be human is to be capable of sin. An animal such as a dog or an ape might return an item to its rightful owner, but only a human can do so both out of a real concept of ownership and a sense of duty toward what is right. By the same token, an animal can take an item from its owner, but only a human can steal it. One can easily construe this, if brought to an extreme, as meaning that sin in fact originates humanity. Just as in the Genesis narrative the man and woman only become conscious of their nakedness by means of their sin, humanity as a moral being is only identifiable within the history of evolution where there exists the consciousness of having already sinned.

In this vein, the American Protestant philosopher John Fiske (1842–1901) argues in 1899 that the Genesis fall narrative actually reveals the necessity of evil within God's good plan for the world. "The rise from a bestial to a moral plane of existence," he writes, "involves the acquirement of the knowledge of good and evil."[31] Without justifying any evils in particular, Fiske insists that evil as such is necessary for the production of the good.[32]

Jonas remains indispensable: *The Gnostic Religion: The Message of the Alien God and the Beginnings of Christianity*, 3rd ed. (Boston: Beacon, 2001).
30   O'Regan, *Heterodox Hegel*, 164; see also 154–69.
31   John Fiske, *Through Nature to God* (Boston: Houghton Mifflin, 1899), 52.
32   Fiske, 25.

While Fiske, like Hick, sees evil as necessary for our moral and spiritual development, he goes further than Hick by treating evil as having a real, positive function. In Fiske's view, because death is an indispensable element of the process of natural selection, some degree of suffering is necessary for humanity to find happiness. Therefore, the difference between good and evil is merely relative: "Moral evil is simply the characteristic of the lower state of living as looked at from the higher state."[33] In other words, evil is merely the unfinished good, the good in process, or to use a phrase from the atheist postmodern philosopher Slavoj Žižek, "Evil is Good itself 'in its becoming.'"[34]

Gregory Peterson provides a straightforward example of a contemporary inverted fall. In a 2004 essay, he describes original sin as "falling up"; it symbolizes the necessary but unfortunate development of human freedom, which unavoidably produces sin in the world. Our psychological complexity provides us with increased freedom, but this freedom remains incomplete and therefore prone to following problematic biological and cultural proclivities.[35] In order to emphasize the naturalness of this "fallenness," Peterson expands the term "sin" to include any and all suffering in the natural world. In this sense, the entire world is both sinful and fallen; humanity only experiences this continual falling (upward) in a more intense manner.

Peterson's attribution of degrees of suffering to all creatures hints at his reliance on process theology. The rise of process theology has contributed substantially to the propagation of the inverted fall narrative today.[36] After all, process theology closely resembles Hegel, and Hegel relies heavily upon an inverted fall. Though we cannot explore these connections fully here, suffice it to say that they are far from incidental. The inverted fall makes evil intrinsically productive and in that sense both explained and explanatory; evil must happen for the sake of the good, and good is thus the end that justifies evil as a means. Such an approach is vital to Hegelian dialectic precisely because the world serves as the ultimate explanation for God. Evil is good-in-process precisely because the world is God-in-process.

---

33 Fiske, 54.

34 Slavoj Žižek, *The Indivisible Remainder: An Essay on Schelling and Related Matters* (London: Verso, 1996), 113.

35 Gregory R. Peterson, "Falling Up: Evolution and Original Sin," in *Evolution and Ethics: Human Morality in Biological and Religious Perspective*, ed. Philip Clayton and Jeffrey Schloss (Grand Rapids, MI: Eerdmans, 2004), 284–85.

36 See for example Birch and Cobb, *Liberation of Life*, 120–21.

Selfish Gene Theology

Though not always incompatible with the other two narratives, the third sees the fall not as something belonging to humanity in particular but as part of nature as such. In other words, it is not a characteristic of human nature but of nature broadly speaking. Natural selection inherently involves competition, violence, and selfishness. Thus, the species or natures that it produces are predisposed to such behaviors. This predisposition is an important element of the natural struggle for existence among species, populations, groups, or individuals. Yet among humans, these deeply ingrained animal tendencies toward competition and violence take on negative moral significance and lead to specific behaviors that run contrary to the greater potential posed by our unique intelligence and broader capacity for social unity. Human nature is competitive and violent because it developed within an evolutionary process that relies upon violence and competition in order to further the development of organisms.

This narrative stems from theological readings of Richard Dawkins and Edward O. Wilson, the fathers of sociobiology, a controversial field of theoretical science.[37] In his provocatively titled book *The Selfish Gene*, Dawkins argues that natural selection occurs first and foremost on the level of genes rather than organisms. Evolution puts genes in competition with one another. This favors competitive traits, which eventually leads to competitive organisms. While genes are not selfish in a moral sense, they are inherently shaped by mutual competition in a manner that intrinsically favors the propagation of successful DNA rather than the prospering of successful organisms. In brief, a koala's genetic adaptation that allows it to eat eucalyptus did not develop primarily for the sake of the koala; rather, the koala developed for the sake of propagating this gene.[38] Organisms change over successive generations, but a successful

---

37 Richard Dawkins, *The Selfish Gene*, 30th anniv. ed. (Oxford: Oxford University Press, 2006); Edward O. Wilson, *Sociobiology: The New Synthesis* (Cambridge, MA: Harvard University Press, 2000). On the differences between the two, see Celia Deane-Drummond, *Christ and Evolution: Wonder and Wisdom* (Minneapolis, MN: Fortress Press, 2009), 62.

38 On the genetics of the koala's digestive system see Johnson, Rebecca N., Denis O'Meally, Zhiliang Chen, et al., "Adaptation and Conservation Insights from the Koala Genome," *Nature Genetics* 50, no. 8 (August 2018): 1104; Chen, Jiayan, Weijie Lv, Xueli Zhang, et al., "Animal Age Affects the Gut Microbiota and Immune System in Captive Koalas (Phascolarctos cinereus)," *Microbiology Spectrum* 11, no. 1 (January 5, 2023): 2.

gene remains the same. As Wilson says, "The organism is only DNA's way of making more DNA."[39]

This is not to say that evolution excludes any kind of altruism. The selfish gene view allows for altruistic behavior but insists that even this traces back to genetic competition. Dawkins rejects, for example, the typical view that individual organisms may act altruistically for the sake of the propagation of the species. Self-sacrificial behavior would really act in favor of those organisms who eschew altruistic behavior and favor their own propagation above that of their peers. Instead, Dawkins argues that organisms sometimes appear to act altruistically, but only so long as it benefits the propagation of their own genes. The self-interest of genes is therefore absolute. The apparent altruism of organisms is merely another vehicle by which particular genes promote themselves above others.

Dawkins's work is rabidly atheistic, and thus theodicy is not his concern. For him to paint selfishness as a fundamental principle of biological reality is merely to state what he sees to be empirical fact. Nevertheless, while he does see evolutionary selfishness as the basis of selfish moral behavior, Dawkins does not subscribe to strict biological determinism. He takes for granted that it is possible for us to resist the selfish urges of our DNA and to rise above selfish behavior through the development of culture. For better or for worse, DNA contributes to our behavior, but it does not force it.

The selfish gene approach has been amply criticized from within biology, and this is in no small part due to Dawkins's argumentative attitude and cavalier use of metaphors.[40] The assertions of selfish gene theory are not easy to dismiss, however, and to the extent to which they often invite philosophical questions, it is only natural that they should come to bear upon theology as well.

Accordingly, several evolutionary theologians utilize selfish gene theory as the basis of their own theological narrative. Catholic examples include Daryl Domning, Jerry Korsmeyer, Nicholas Olkovich, and Jack Mahoney; Protestant examples include Philip Hefner, Gregory Peterson, and Patricia A. Williams.[41] For convenience, we can refer to those who

---

39 Wilson, *Sociobiology*, 3.

40 For more on this, see Evelyn Fox Keller, *The Century of the Gene* (Cambridge, MA: Harvard University Press, 2002); Midgley, *Evolution as a Religion*, xi.

41 Daryl P. Domning, *Original Selfishness: Original Sin and Evil in the Light of Evolution* (Aldershot, England: Ashgate, 2006); Korsmeyer, *Evolution*; Nicholas Olkovich, "Reinterpreting Original Sin: Integrating Insights from Sociology and the Evolutionary Sciences," *Heythrop Journal* 54, no. 5 (2013): 715–31; John Mahoney, *Christianity in*

espouse this third narrative as "selfish gene theologians." However, such a moniker should not be taken too seriously. These figures do not all teach the same thing, they do not all agree with Dawkins, and in particular they vary on the extent to which they actually emphasize the dominance of selfishness.

While selfish gene theology often includes the gradual ascent and/or the inverted fall, the hallmark of selfish gene theology is the claim that original sin represents the aftereffect of the inherent selfishness of the evolutionary process itself. We struggle against self-interested tendencies that evolved as a result of the competitive matrix of evolution.[42] It is not simply that our nature is violent; rather, the very process of evolution is predisposed toward producing competitive organisms.[43] Competition shapes nature and propels selfish and violent behavior. This led to such behaviors becoming enmeshed within the culture of our prehuman ancestors, who passed it on to us.[44] In this way, we are selfish both by nature and by nurture. Biology and sociology conspire. Surrounded by biological and cultural propensities for sin, we require redemption at the deepest level possible.

While Gregory Peterson sees this inherited situation itself as sin, others are more careful to maintain a sense of sin as a voluntary effort.[45] For Philip Hefner, the disjunction between our genetic predispositions and our cultural heritage presents a challenge to our development and makes us vulnerable and fallible as we struggle by trial and error to synchronize these biological and cultural cues within the demands of present-day experience. Our free-will must operate within a situation marked by such vulnerability and fallibility, and it is because of this that we tend to sin. Nevertheless, vulnerability and fallibility are so intrinsic to the human person that Hefner considers them to belong to the very image of God (Gen 1:26–27).[46]

Evolution: An Exploration (Washington, DC: Georgetown University Press, 2011); Philip J. Hefner, "Sociobiology, Ethics, and Theology," *Zygon* 19, no. 2 (1984): 185–207; *Human Factor*; Patricia A. Williams, *Doing Without Adam and Eve: Sociobiology and Original Sin* (Minneapolis, MN: Fortress Press, 2001). For an important examination and critique of such views, see Ted Peters, "The Evolution of Evil," in *The Evolution of Evil*, ed. Gaymon Bennett et al. (Göttingen: Vandenhoeck and Ruprecht, 2008), 19–52.

42 Korsmeyer, *Evolution*, 122.
43 Patricia A. Williams, "Sociobiology and Original Sin," *Zygon* 35, no. 4 (2000): 791.
44 Domning, *Original Selfishness*, 178; Korsmeyer, *Evolution*, 122–25; Hefner, *Human Factor*, 132.
45 Peterson, "Falling Up," 284.
46 Hefner, *Human Factor*, 240–41.

Comparison and Critique

Notwithstanding their differences, each of these three narratives reconsiders the doctrine of original sin for the same key reason: the problem of evil. As we have seen, theodicy became a central concern with the rise of modernity and only grew in importance alongside evolutionary theology in general and process theology in particular. These narratives have evolved for the purpose of explaining how a good and loving God could have created a world rife with suffering and death. Such represents their chief benefit but also their weakness, for it remains debatable whether any of these narratives really produces a consistent and worthwhile explanation for the origin of evil.

Premodern theologians typically only address one of two dimensions of the problem of evil. They provide many explanations for *moral evil*—the evils that humans commit as free actors—while they set aside nonmoral or *natural evil* as a mystery of the divine will. God permits animal violence and earthquakes alike without having to justify them to us.[47] Animals suffer and die, and this is simply part of God's design for the world. Historically, this lack of a strong explanation for natural evil became all the more stark in view of natural selection, which makes death a necessary mechanism by which the environment selects against certain organisms or adaptations in favor of others. An awareness of this helps us to recognize the exponential magnitude of animal suffering in the world. In order for a beneficial adaptation to become dominant within a population, thousands of less fit organisms must die. In short, traditional approaches to theodicy typically provide no explanation for what John Haught terms God's "wasteful" creativity: the necessity of pain and suffering for the sake of evolutionary adaptation.[48]

Accordingly, the principal contribution of these three evolutionary narratives of original sin is their ability to unify the two dimensions of the problem of evil.[49] Eschewing the traditional fall narrative, they see moral evil as an extension of natural evil. This allows them to view original sin as a situation that encompasses both forms of evil at the same time. One way or another, and to varying degrees, this leads them to see moral evil as either unavoidable

---

47 Thomas Aquinas, *Summa theologiae* I, q. 6, a. 1 ad 2. Note that in the modern era, conversely, it became common among many Protestants to think of animal death as a consequence of original sin. See Simpson, "Darwin," 375.

48 Haught, *Making Sense*, 38.

49 Williams, *Doing Without*, 169; Domning, *Original Selfishness*, 157. See Peters, "Evolution of Evil," 21.

or necessary for the development of the good. Put another way, the existence of human, moral evil is tied to or guaranteed by the existence of natural evil. Humans sin precisely because suffering and death are inextricable aspects of the natural world.

Following this logic, all three narratives make human sin a matter of cosmic immaturity. The gradual ascent narrative downplays sin as incidental—mere growing pains already on the way out. Although it often depicts sin as a passive state of imperfection, it can also speak of it as an active stagnation, a refusal to move forward.[50] The inverted fall narrative takes sin more seriously, but only because it sees sin as a necessary component in the production of the good. Evil is an inevitable and necessary step on the way to the realization of evolution's higher goal. Selfish gene theology can agree with both points, but more specifically it sees sin as an echo of the evolutionary process. Because evolution is inherently competitive, because it operates on the principle of selfishness, it is only logical that it would produce selfish organisms. In all cases, human sin is thus the extent to which this immature precondition remains dominant despite being ill-suited to the moral and social demands of modern human society.[51] Blatant self-interest may allow natural selection to produce better capybaras, but acting on such an impulse becomes a culpable offense in light of the higher concerns of human existence.[52]

The solution to the problem of moral evil, therefore, becomes a matter of maturation. Something about humanity must grow to overcome the tainted inheritance of our evolution. These narratives sometimes point especially to culture as the key site of further development. While we are prone to sin by nature, they reason that freedom or culture is able to reject such impulses and to give pride of place to higher social, moral, and spiritual concerns. Biological evolution gave birth to us, but spiritual evolution must become our new mother and true educator. Such development is not something we passively receive. It is rather an active task that we undertake in free cooperation with God's plan. In fact, for those influenced by process theology, salvation is the ongoing process whereby we collaborate "with God as co-creators of the universe."[53]

---

50 See for example Ansfridus Hulsbosch, *God in Creation and Evolution*, trans. Martin Versfeld (New York: Sheed and Ward, 1966), 43.
51 Domning, *Original Selfishness*, 151.
52 Korsmeyer, *Evolution*, 122–23.
53 Domning, *Original Selfishness*, 151–52.

While such tidy and simple solutions are attractive, they raise several problems. For lack of space, we can only mention here a few key critiques. These critiques do not apply equally to all advocates of these views. In fact, some versions of these narratives are stronger than others because they are better able to respond to some of these critiques.

First, apart from a few notable exceptions, these views dramatically downplay the significance of Jesus Christ. Christ is not necessary for salvation, but perhaps serves as an added bonus. If anything, he does not save us by his cross, which was simply the inevitable result of his rabble-rousing and not some transcendent and efficacious sacrifice.[54] Rather, Christ serves as a model of imitation as one who overcame the impediment of original sin and fully embraced the love of God. Grace refers only to how we encounter or learn about God in and through nature, including our own natural capacity for freedom.[55] For Williams, this makes grace our natural human tendency for altruistic behaviors toward our kin, which she sees as the basis of love of God and love of neighbor.[56]

To be clear, these are Pelagian positions. Without going into detail about the historical debate, Pelagius insisted that grace was not necessary for salvation but only a helpful extra. In connection with this, he tended not to think of grace as an active power. At times he identified grace with our natural, God-given abilities, and at other times he saw it as a body of good teaching. God thus provided grace first in the act of creating us and second in sending Christ to teach us by word and example. We became sinners by imitating bad examples, and we find salvation by imitating the good example of Jesus. Because Pelagius believed that the will was unencumbered by original sin, it followed that knowledge alone was sufficient to lead us to good deeds. If we know what is right from the example of Jesus Christ, then logically we will do what is right.

A full array of anti-Pelagian critiques could fill its own volume, so suffice it to say here that Pelagius was condemned for good reasons.[57] Not only is this *not* what the bulk of the Christian tradition means when it talks about grace and salvation, but Pelagianism provides a message that is ultimately

---

54 Domning, 152; Williams, *Doing Without*, 195–97.
55 Hefner, *Human Factor*, 42, 45, 61–62.
56 Williams, "Sociobiology," 807.
57 Augustine lays out and criticizes Pelagius's views in "The Grace of Christ and Original Sin" in *Answer to the Pelagians*, trans. Roland J. Teske (Hyde Park, NY: New City, 1997), 384–463.

ineffectual and impotent. Those who struggle against sin are left with no real help. The truth is that far too often we already know what is right but we do what is wrong anyway. Hence Paul says, "What I do, I do not understand. For I do not do what I want, but I do what I hate" (Rom 7:15). A Pelagian perspective insists on overcoming sin by sheer willpower because it sees salvation more as one's own action than as a divine gift.

Second, despite their best intentions, all of these narratives to varying degrees risk exonerating humanity for its evils. This is a natural consequence of applying a singular model to unify natural evil and moral evil. If the existence of moral evil is tied to natural evil, if humans are bound to sin because of the underlying influence of evolutionary selfishness, then on what basis can we critique those who commit evil?[58] Put another way, if we are driven to sin because of our immature evolutionary nature, is not evolution or nature ultimately to blame? Of course the goal of these views is to exonerate God, but the more that we project sin onto our evolutionary past, the more we diminish our own responsibility, the easier it becomes to blame God after all.

This problem is deeply related to a tendency among these narratives to downplay the real gravity of sin. As we saw, Teilhard was especially notable for this tendency, but other even less optimistic evolutionary theologians tend in this same direction. Of particular note is the selfish gene theologian Patricia Williams, who argues that, as a rule, whenever evil enters into the world a greater proportion of good must enter at the same time. "Because good and evil are conjoined," she writes, "evil is redeemed as it arises."[59] For example, evil entered the world when creatures evolved the ability to feel pain. However, this is justified or redeemed, she argues, because this development is tied to the greater good of sentience. In sum, evil is essentially convertible with the good. Even human death, she insists, is validated by the process of decomposition. The evil of mortality serves the greater good when we become food for the worms.[60]

---

58 Deane-Drummond, *Christ and Evolution*, 86. See also Cynthia S. W. Crysdale and Neil Ormerod, *Creator God, Evolving World* (Minneapolis, MN: Fortress Press, 2013), 91–95.

59 Patricia A. Williams, "How Evil Entered the World: An Exploration Through Deep Time," in *The Evolution of Evil*, ed. Gaymon Bennett et al. (Göttingen: Vandenhoeck and Ruprecht, 2008), 205, 217.

60 Williams, *Doing Without*, 178; "How Evil," 214; cf. Domning, *Original Selfishness*, 51.

Such a quantitative reckoning of good and evil is fundamentally flawed, since it assumes that good and evil have some common exchange value (e.g., a pound of good = a pound of evil). It seeks to safeguard God's justice on mere quantitative terms, as though God is good so long as the world contains a few extra grams of good. This is, in other words, a kind of economic logic: the end justifies the means so long as things are a little better after than before. Evil is an investment that God makes hoping for good returns, and the price to be paid is validated if God earns more than God spends.

In the end, one must wonder whether such platitudes would have any meaning to those who are actually undergoing real, serious, painful, and disastrous suffering due to the sin of others. To someone who has received an unjust proportion of the world's evil, it may offer no comfort to think that somewhere *else* in the world there must be a greater portion of good. Likewise, a murder victim's family might not find any comfort in knowing that their daughter's corpse is being recycled by microbes and worms.

Minimizing evil is problematic, moreover, precisely because our real capability for evil is dependent upon our more fundamental capability for good. The traditional doctrine of original sin holds that we are able to dramatically shape the course of history—that our first parents actually marred our very nature through their corrupt decision—precisely because freedom is inherently consequential. Our potential to bring about world-shaking evil is only a lesser expression of freedom's capability to change the world for the better. If we cannot damn ourselves, then neither can we save anything.

Third, one must consider the political implications of such views. In fact, Williams's theodicy actually predates Darwinian evolution, as much the same arguments can be found in the work of Thomas Malthus. For Malthus, evil is necessarily a part of God's creative plan. God could not achieve the supreme goal of the formation of mind out of matter without some admixture of evil. Evil—both in terms of natural suffering and human sin—serves as a necessary impetus toward action. Without suffering and death, humans would give in to laziness and vice. Nevertheless, this amounts to good news:

> The partial pain . . . is but as the dust of the balance in comparison of the happiness that is communicated; and we have every reason to think, that there is no more evil in the world, than what is absolutely necessary as one of the ingredients in the mighty process.[61]

---

61 Malthus, *First Essay on Population, 1798*, 391.

Evil is justified both *ultimately* (the end justifies the means) and *quantitatively* (there is always more good than evil in the world).

Malthus's theology is inextricably bound up with his political convictions. He pens his ideas with the chief purpose of destroying the growing store of liberal enthusiasm for government-funded welfare programs. A skilled rhetorician, Malthus frames his argument in scientific terms, seeking to show such attempts at social improvement to be against nature itself. "Rather than an enraged defender of the social order," David McNally notes, "Malthus thus assumed the guise of a dispassionate investigator of the natural order."[62] Accordingly, while Malthus recognizes that suffering falls disproportionately upon the poor, he sees this as a just arrangement meant to counteract their natural laziness. Any attempt on the part of the wealthy to nullify this disparity serves only as a model of hubris condemned to eventual failure. After all, he reasons, without sufficient hardship to cull their numbers, the poor will breed without limit, leading to widespread shortages and a preponderance of suffering for rich and poor alike.

In fact, the economic structure of Malthus's theodicy is tailored to fit his political convictions. Evil is, in essence, quantifiable. While Malthus cannot provide a specific quantity, he attests to the mathematical expression: *good in the world > evil in the world*. The specific quantity of evil, moreover, is strictly justifiable by the good sought by the divine plan; it is, in other words, a fair and appropriate price. Just as laissez-faire capitalism believes that the market will regulate itself, so also does Malthus believe that the balance of good and evil is naturally regulated by the threat of overpopulation. The suffering and death of the poor represent a partial, necessary evil. By undermining any attempts to build a perfect world—to go against the natural order—they serve to protect the world from the far greater evil of universal suffering. At the same time, they also produce hard work and strong character among those who survive.

In short, Malthus's popularity among early eighteenth-century Whigs and twenty-first-century Libertarians makes perfect sense. By both inscribing economic inequality into the natural order and at the same time representing wealth and poverty as tied to the individual's moral character, Malthus legitimates the *status quo* and defends the intrinsic justice of the free market.

---

62 David McNally, "Political Economy to the Fore: Burke, Malthus and the Whig Response to Popular Radicalism in the Age of the French Revolution," *History of Political Thought* 21, no. 3 (2000): 436.

Yet not only must one be suspicious of his claim that this represents the very truth of Christian charity, but one must also wonder whether such views can have purchase within the vast world that suffers beneath comfortable, first world, upper-middle-class existence.

Our final critique is the most important. The more that these views provide an *explanation* for the problem of evil, the more they tend to undermine any real *solution* to this same problem. Process-based approaches, for example, insist that evil must exist because God is incapable of removing it.[63] In fact, God needs evil in order to become good in any definitive sense. God must become subject to the inevitable happenstances of finite existence, must suffer such experiences in order to find fulfillment. Accordingly, God does not and cannot answer prayers.[64] Recall that from a Whiteheadian perspective, divine providence means that God serves as a lure for feeling. At its maximum, this means that God draws the world's process forward in ways that stimulate meaningfulness. It in no way means that God actually provides any real, tangible help to those in need.

Given such a useless God, it is little wonder that many theologians turn to the idea of divine suffering. If God cannot help us in our need, then at least God can commiserate alongside us. Yet in the end, such a God is constitutively incapable of saving us. How can a God who is in need of salvation give what God's own self does not have? On what basis, then, can we hope for any ultimate solution to evil?

In this way, many evolutionary views on original sin constitutively lack a robust theology of grace. Either they adopt the weak God of process theology or the self-limiting God of John Hick, or else they cast salvation in terms of culture or imitation. We are called to rise above our evolutionary immaturity by, for example, imitating the pattern of Jesus's life.[65] Certainly the imitation of Christ is a central element of Christian tradition, but if this imitation is merely the modeling of an external pattern (without the infusion of divine grace), then how can such an extrinsic influence ever overcome the deep-seated and intrinsic problem of original sin? If our very nature

---

63 See David Ray Griffin, John B. Cobb Jr., and Clark H. Pinnock, eds., *Searching for an Adequate God: A Dialogue between Process and Free Will Theists* (Grand Rapids, MI: Eerdmans, 2000), xi.

64 Bruteau, *God's Ecstasy*, 176–77.

65 Domning, *Original Selfishness*, 152.

is constituted by the process of evolution, then how could mere imitation enable our efforts to overcome this same nature? In short, such hope in the triumph of human goodness is wishful thinking. If on the contrary we take evolution seriously, recognizing that it has shaped us on the most intimate level—if we are shaped by evolutionary selfishness to the very core—then how can we produce an action that goes against everything that we are? It is hard enough for a smoker to quit the habit; how can a human quit everything that makes one human?

What evolutionary theology needs is a deeper understanding of divine grace. There are hints in this direction. Jack Mahoney, for example, attempts to solve these problems by appealing to the image of God in the human person as the source of a divine altruism.[66] While he attempts to show that this is compatible with his selfish gene premise, his insistence on the divinity of altruism risks blurring the distinction between nature and grace. Rather than attempting to make human nature inherently altruistic, a thoroughgoing theology of grace can argue that God brings out what is positive about human nature. Yet once again, this requires an understanding of salvation as pure gift and a rejection of any Pelagian notion of spiritual maturation through effort, culture, or knowledge.

As we have seen, these evolutionary narratives of original sin begin with a denial of the historical fall in order to propose a new and comprehensive answer to the problem of evil. They presume that original sin is first and foremost a theodicy, and thus they make theodicy the be-all and end-all of this doctrine. Yet in doing so, they typically exonerate God for evil only by projecting evil into the evolutionary past. This results in an explanation for evil's existence, but little in the way of a satisfying solution. In fact, in the end, they often effectively make God responsible for evil but excuse God on the premise that God could not have done otherwise.

While such narratives are certainly popular among evolutionary theologians, they do not represent the only means of reading original sin through an evolutionary lens. We turn now to two key exceptions from the Catholic tradition who, contrary to others, insist that original sin actually *happened* and that this, in fact, reveals the real significance of evolution for original sin.

---

66 Mahoney, *Christianity in Evolution*, 44.

EVOLUTION AND THE HISTORICITY OF ORIGINAL SIN

The three narratives mentioned above ultimately project the origin of evil back onto God. Evil is merely the incompleteness or immaturity of an evolutionary world or it is the experience of pain that evolves alongside positive experiences. God must create such a world either because God is incapable of destroying evil or because the end that God seeks requires such a means. God is responsible for evil, but for one reason or another God is not *indictable* for evil.

Yet original sin was never meant to be a complete theodicy, and as a doctrine it does not explain the existence of natural evil. What it does explain is moral evil, and it provides a better explanation for that because it includes a *free-will theodicy*: a claim that human evil is entirely the product of humanity's free decision and not something willed by God or necessitated by some higher end. Certainly the *possibility* of evil is a necessary byproduct of the *possibility* of good, but the *actuality* of evil is a mere happenstance and a product of real persons' real decisions.

Does evolution inherently point us to these three narratives? If we accept that humans have evolved, and that the science of evolution has something positive to contribute to theology, then must we also accept that original sin is merely a theodicy, which is better represented by a gradual ascent, an inverted fall, or the intrinsic selfishness of evolution?

In fact, there are other options, and two figures in particular point us in a fruitful direction: Karl Rahner and Raymund Schwager. They stand apart for being both innovative and traditional at the same time. Although Rahner has often been cited in support of the above views, in reality his theology of original sin aligns more closely with that of Augustine. This is because both Rahner and Schwager begin with the vital premise that the historical deed of Adam and Eve is, somehow or other, a reality. While evolution does play a role in how original sin impacts our lives, original sin is not a function of nature or evolution itself. Rather, the result of the free deed of our first parents redounds unto us by way of historical influence.

This historical influence is what I term the *historicity* of original sin. Historicity here means not only that original sin happened at some time in the past but also that it continues to have an impact on subsequent history precisely because it is a historical event. Any genuinely real historical event has historicity because, by entering into history, it shapes what follows after it. In this way, the COVID-19 pandemic not only happened, but continues to influence global attitudes, actions, and culture.

The historicity of original sin impacts us on the deepest level possible. As Schwager explains, the science of evolution proves that original sin was right all along in blurring the distinction between nature and freedom.[67] As it turns out, free, unscripted historical actions can indeed shape our very nature. Based on the work of the physicist and biologist Carsten Bresch, Schwager argues in particular that violent competition likely contributed to the development of the human brain.[68] The decisions that early humans made in regard to friendship or violence, mating or murder, unity or difference all contributed to the genetic makeup of our species. As a result, sin and evolution can no longer be neatly disentangled.[69]

Celia Deane-Drummond and Benno van den Toren approach original sin from a similar angle, utilizing insights from the scientific theory of niche construction. The point of this theory is that, while it is true that organisms are shaped by their environment through natural selection, there are also important ways that organisms change their environments as well. One example is plant–soil feedback. The dropped leaves of a tree can affect the nutrients and acidity of the soil around it in ways that favor the prospering of the tree and its species. Through such influences over long periods of time, populations and environments evolve in close conjunction with one another. "Species do not only adapt to different environments," explains Van den Toren, "but they also adapt environments to themselves."[70] Because humans have an advanced capacity for cultural development, human culture plays a strong role in the kind of evolutionary feedback that constructs our own particular niche. Such complex cultural feedback pushes back against a simplistic identification of

---

67 Raymund Schwager, *Banished from Eden: Original Sin and Evolutionary Theory in the Drama of Salvation*, trans. James Williams (Leominster, UK: Gracewing, 2006), 4–7; see also 54, 105–106, 117.

68 Carsten Bresch, *Zwischenstufe Leben: Evolution ohne Ziel?* (Frankfurt: Fischer, 1979), 196–202; Schwager, *Banished*, 55.

69 Karl Rahner, *Meditations on Priestly Life*, trans. Edward Quinn (London: Sheed and Ward, 1973), 49.

70 Benno van den Toren, "Original Sin and the Coevolution of Nature and Culture," in *Finding Ourselves after Darwin: Conversations on the Image of God, Original Sin, and the Problem of Evil*, ed. Stanley P. Rosenberg et al. (Grand Rapids, MI: Baker Academic, 2018), 177.

original sin with the selfishness of nature. Rather, the freedom of human culture plays an intimate role in shaping who we are, for better or for worse.

For Rahner, Schwager, and Deane-Drummond, if human culture can and does shape human evolution, then *history* is the real vehicle of original sin. They are flexible on the precise details, but all affirm that *sin happened* at the origin of human history and that this began the inherited situation of original sin that we experience today.[71] Our first parents were not essentially different from us. Rather, the reason their misdeed continues to impact our situation is precisely because every sin has historical consequences. We make decisions today that impact the freedom of those who come after us. Our first parents were in a relatively unique position because they stood at the beginning of the race, a point that unites all who come after.

As we discussed in chapter 4, Rahner recognizes that the human population may never have been as small as two. The propagation of original sin is premised on human unity, and this unity includes but goes beyond biological unity. The first human population was united not only by origin and procreation but also by social relationships and their grace-given destiny toward fulfillment in Jesus Christ.[72] On the basis of such unity, the free deed of even a portion of these original humans can dramatically reshape humanity's future, including its evolutionary trajectory and its relationship with the Trinity.

After all, according to such a perspective, the unity of the human race is both a premise and a task.[73] The grace of original justice was meant to deepen our basic unity with one another by incorporating us into a deep relationship with God. According to God's original design, human unity and biological descent *should have* been the means for transmitting the grace of original justice. Mere membership in the human race should have meant participation in its unique relationship with the Trinity. *Grace should have belonged to human beings as though naturally.*[74] Due to the advent of sin, humans are not

---

71 Celia Deane-Drummond, "In Adam All Die?: Questions at the Boundary of Niche Construction, Community Evolution, and Original Sin," in *Evolution and the Fall*, ed. William T. Cavanaugh and James K. A. Smith (Grand Rapids, MI: Eerdmans, 2017), 27, 35–36. See also Gijsbert van den Brink, "Questions, Challenges, and Concerns for Original Sin," in *Finding Ourselves after Darwin*, ed. Stanley P. Rosenberg et al. (Grand Rapids, MI: Baker Academic, 2018), 117–29.

72 Rahner, "Evolution and Original Sin," 67; "Exkurs: Erbsünde und Monogenismus"; see "Unity," 156.

73 Rahner, "Unity," 155.

74 *Foundations of Christian Faith: An Introduction to the Idea of Christianity*, trans. William V. Dych (New York: Crossroad, 1982), 113; "Exkurs: Erbsünde und

by nature the conduits of God's love that their Creator intended. *Peccatum originale originatum* (originated original sin) is thus our situation of having lost, through the deed of our first parents, that initial and inborn state of grace.

Whereas human unity was meant to be the vehicle of grace, it has instead become a vehicle of sin. By choosing sin and rejecting grace, our first parents not only frustrated God's plan but chose thereby to unite under a contrary principle. Humans became coconspirators. Schwager's use of René Girard (1923–2015) is helpful here because Girard emphasizes how human communities tend to unite by casting blame upon one or more persons who serve as scapegoats. In other words, the archetypal sin from a Girardian perspective is the sinful unity of a lynch mob. Yet there is no honor among thieves, and such attempts can only maintain a semblance of unity for so long. Eventually, violence breaks out once again. More sin is required in order to maintain periodic bouts of peace.

We cannot in this short space fully explore the insights of these thinkers, but it is important to recognize that their approach is not intended to provide one complete and simple answer. In contrast to the above narratives, the historicity approach does not unite natural evil and moral evil into one unified theodicy. Moreover, Rahner in particular deals with original sin from several distinct angles, all of which must be brought together in order to form a complete picture of the doctrine.

One key angle of Rahner's approach touches further on the topic of sin's historicity by arguing that original sin entails the *codetermination of freedom by the guilt of others*. To put it simply, this means that every time we make a free decision, we have no choice but to make it within a world and a milieu that is already shaped by the free decisions of others—for good or bad. Rahner gives the paradigmatic example of the plight of exploited banana pickers. When we buy food at a supermarket, we may be implicitly supporting an unjust system of labor somewhere far away without even realizing it.[75] Put another way, the history of evil breeds more evil and even undermines our best attempts to do otherwise.

Importantly, Rahner's approach is helpful for considering how original sin relates to what liberation theologians term *social sin*—structures within

---

Monogenismus"; cf. "Sin of Adam," 259; see Nikolaus Wandinger, *Die Sündenlehre als Schlüssel zum Menschen: Impulse K. Rahners und R. Schwagers zu einer Heuristik theologischer Anthropologie* (Muenster: Lit, 2003), 81–82.

75 Rahner, *Foundations*, 110–11.

culture, politics, and society that enshrine evil and unjust situations into the established order. Segregation and apartheid, for example, originated in the actual sinful deeds and attitudes of real persons but became so extensive and deeply established that they transcended the guilt of mere individuals. In such a situation, many people even become unwitting victimizers of others because they simply accept as natural and given a situation that is really the embodiment and crystallization of prior sin. Importantly, social sin cannot be combated solely on an individual level. Real change takes more than the conversion of a few people. Accordingly, social sin requires *both* individual conversion and social (cultural and political) action toward a brighter future.

The problem of social sin highlights the fact that sin too has a history. Sin undergoes a kind of evolution. The situation of sin that people faced a thousand years ago is not exactly the same as the situation that we face today. Each era invents new forms of injustice and new weapons of violence, and thus each era must face the problem of sin anew.

Rahner, Schwager, and Girard all recognize this problem and point each in his own way to the one and only solution: grace. Just as sin creates a history, so also grace forms its own more primordial, more fundamental history. After all, the history of grace began before the history of sin and continues to this very day as God actively aids us in combating the snowballing forms of sin in our world.

Having explored the past of original sin, it is time to examine the possibilities of the future. Evolutionary theology is founded on the idea that history has a kind of forward trajectory. Accordingly, the idea of the future plays a central theological role. Yet the precise character of the future varies from thinker to thinker. As we shall see, different emphases in regard to the future open up distinct areas for theological exploration.

CHAPTER SIX

Future Horizons

As we have seen, the very idea of evolution has a broad discursive power. Its draw among philosophers and theologians often has less to do with its scientific merits and more to do with its wealth of emotional and intellectual stimulation. In particular, by coordinating the complex history of the world into a narrative of progress, evolutionary schemas provide a compelling and energizing sense of a complete and organized whole.

Yet the power of evolutionism is not exhausted in speaking about the past. The essence of evolutionary theology leads thinkers to consider also the situation of the present and the future toward which it points. As Julian Huxley writes:

> Since evolutionary phenomena (of course including the phenomenon known as man) are processes, they can never be evaluated or even adequately described solely or mainly in terms of their origins: they must be defined by their direction, their inherent possibilities (including of course also their limitations), and their deducible future trends.[1]

Evolutionary forms of theology tend to link their understanding of the past to both an analysis of the present and expectations of the future. Without necessarily rejecting Christianity's traditional hope in an otherworldly salvation, evolutionary theologians seek to envision the future of *this* world as the next chapter in the ongoing drama of evolution.

In fact, the dimension of the future is vital for determining the meaning of the evolutionary whole. Conrad Bonifazi explains:

---

1 From Huxley's introduction to Teilhard de Chardin, *Phenomenon of Man*, 13.

> [Evolutionary forms of thought] need an end-term to supply the element of continuity and to unite series of changes into a single process. Processes themselves translate this end-term into 'something' which is both coming into being and also present throughout the stages of its development. The final phase unites the process and comprehends the whole.[2]

In other words, the coming end shapes the evolutionary process and constitutes it as one interconnected movement toward an ultimate goal. The intimate relationship between past, present, and future forms the basis upon which evolutionary theologians can both suggest a theological praxis (what should we do now?) and believe that the accomplishment of such goals is really and truly possible.

In view of this, the horizon of the future functions in three particular ways, which are emphasized differently by diverse thinkers. First, the envisioned future stands as an ideal or the site of ultimate fulfillment. Teilhard de Chardin's Christocentric concept of *Omega* illustrates this well, since Christ brings an eschatological fulfillment that is both "already" and "not yet." The *Omega* is already present within the world in the incarnation, even though it is also progressively breaking into the world as history reaches out toward ultimate fulfillment. Second, to the extent that evolutionary theologians tend to envision evolution as a genuinely creative—as opposed to deterministic—process, the envisioned future also serves as a horizon of open possibilities. Third, to the extent that such an emphasis on creativity serves to underline the world's genuine autonomy, this future horizon resides immanently within the natural possibilities of the world itself. It represents the outcome, the capstone, and the culmination of the drama of the world's meaningful self-development. Hence, for Whitehead, the evolutionary rise of the world is not due to any external divine coercion but to the logic of creativity that characterizes real being as such. God merely stimulates the world's inner creativity.

Evolutionary theology's tendency to envision future possibilities is built into its very framework. It is integral to its very identity. If the world is evolving, then it is moving or in flux; if its future is an authentic development of the present, then its future must be perceptible from the point of view of the present and its trajectory.

Yet as evolutionary theology gazes into the future, it does not produce homogeneous results. Although many theologians entertain relatively

---

2 Bonifazi, *Soul of the World*, ix.

optimistic views about the future, few match the enthusiasm of Teilhard de Chardin. Even among these optimists, vast differences often separate the various futures they envision. Though pessimism is uncommon among evolutionary theologians, a few do view the future with realistic concern. Some others, such as Celia Deane-Drummond, fall somewhere in the middle. She presents a positive image of humanity's peaceful coexistence with nature not because she thinks such a future must someday come about by necessity, but rather because this vision is meant to inspire fruitful attitudes toward the created world here and now. At the same time, she critiques views that assign an overwhelmingly positive value to technological advances.[3]

Such shades of difference highlight the fact that evolutionary theology often carries immediate moral consequences. Even Teilhard, who believed that the positive future would come about no matter what, still saw his theology as issuing a dramatic call to action. While Deane-Drummond's call to action serves as a critique of present-day complaisance, Philip Hefner's approach incites us to dive bravely but reflectively into the onrushing current of technological progress.[4]

Our goal in this chapter is to consider the various ways in which evolutionary theology unfolds a future for humankind. In what sense does biological evolution lead forward? Will humanity's ongoing development primarily take the form of cultural or moral improvement? Will technology lead evolution to transcend the drama of natural selection and enter into a future determined by humanity's ever-expanding intellectual and spiritual capacities? What relevance does this have for the broader, natural world?

Various forms of evolutionary theology draw distinct conclusions about humanity's evolutionary future. Framing their viewpoints in terms of moral and cultural horizons, technological horizons, and ecological horizons will demonstrate how these various conclusions tie into particular evolutionary approaches. First and foremost, however, we must look at how some evolutionary frameworks utilize starkly different views of history. This attitude toward history does not strictly change the envisioned future, but it does have dramatic consequences for how such futures address us here and now.

---

3 Celia Deane-Drummond, *The Wisdom of the Liminal: Evolution and Other Animals in Human Becoming* (Grand Rapids, MI: Eerdmans, 2014), 316–17; *Christ and Evolution*, 281–82.

4 Philip J. Hefner, *Technology and Human Becoming* (Minneapolis, MN: Fortress Press, 2003), 84.

OPENNESS AND CLOSURE

Much of modern thought is shaped by the idea of progress: a sense of a forward movement in history, culture, knowledge, and even religion. The consequent confidence that some thinkers show in regard to the future can easily evoke the charge of determinism: it can appear that the future is already decided, as though every deed and facet of history is preordained from on high.

Yet while we cannot delve here into a survey of philosophies of history, it must be noted that determinism in the most proper sense is quite rare. Hegel and Whitehead in particular present key examples of how attempts to envision the forward movement of history often seek to reconcile the determinations of the past with the openness of the future. In other words, they seek to show how history can have a real inner direction and movement while still being the product of real people making real, free decisions, which may even on the surface counteract that same sense of progress.

While evolutionary thinkers can vary dramatically in how they envision the future and the movement of progress, they share in general in the attempt to balance openness and closure, freedom and determination. After all, they typically see creativity as essential to the evolutionary process, and creativity is a kind of openness toward unforeseen possibilities. A future horizon, therefore, is not a fully predictable and guaranteed event but rather a free possibility that is to a greater or lesser extent the goal of present-day trends or aspirations.

Hegel and Whitehead balance openness and closure by way of dialectic. As we saw in chapter 3, dialectic integrates opposites by way of their opposition. Put simply then, these thinkers see history as *both* deterministic *and* free at the same time. In fact, the real movement of history occurs through the clash of these two most intimate factors. As a result, real history becomes the free sublimation of determinism by way of freedom. For Hegel, freedom is the meaning and goal of history, and all of history is ultimately ordered toward the development of this goal.[5]

In order to make sense of history in this way, Hegel plays on multiple distinct meanings of freedom. Considered as the goal of history, freedom is Spirit. Morally, this would look like a situation where every single individual freely understands, accepts, and owns one's own moral obligations. History reaches out toward freedom as Spirit, but it is also driven from within by another

---

5 Hegel, *Lectures on the Philosophy of History*, 19.

kind of freedom: freedom as open indeterminacy or chaos. This means that even that which seems chaotic, arbitrary, meaningless, and disastrous within history is mysteriously part of a higher purpose or the providence of divine reason. For example, the fall of the Roman empire may have been problematic in itself, but in view of the whole, it was a necessary and integral part of the world's onward development.

In this way, Hegel and Whitehead see chaos or indeterminacy as fundamental to the ongoing history of the world. Hegel explains this using the analogy of a house.[6] The house's site plan or blueprint organizes raw materials such as iron, wood, and stone into a rational structure. Using these, however, requires the action of chaotic elements such as wind, fire, and water, which shape the materials to fit the needs of the design, purifying the metal, growing the wood, and shaping the stone. Yet in the end, the house—a rational, orderly structure—serves precisely as a defense against these same chaotic elements: the house stands against wind, fire, and rain. In effect, the finished house is a synthesis of order and chaos, necessity and freedom, and as such it always contains something that exceeds the blueprint's original intention. Nevertheless—following this analogy—because it contains the chaos of fire just underneath the surface, the wooden house remains forever liable to burst out into flames.

This hidden, chaotic underlayer serves as the basis of the ongoing dialectical movement. Hegel sees history as the process whereby the underlying and apparently suppressed element of chaos or indeterminacy breaks through in order to challenge and destabilize the fragile order. Chaos or indeterminacy is the openness of unrestricted possibility; it is "the negative principle for which nothing is established or absolutely sacred, but which can risk and endure the loss of anything and everything."[7] The negative shakes our expectations, undermines our comfortable *status quo*, and forces the world to reconfigure itself. Only through such struggles can the world achieve the unity of positive and negative: e.g., the reconciliation of freedom and order, individual and state.

Accordingly, history is a narrative of the development of freedom from chaos to authentic subjectivity. Chaos is freedom only in an abstract and arbitrary sense; it is pure randomness, indeterminacy, and lack of meaning. True freedom, in contrast, can only exist through the perfectly and dialectically

---

6 Hegel, *Lectures on the Philosophy of History*, 25.

7 Hegel, *Phenomenology of Spirit*, §387.

integrated society—the fully realized state. In brief, individuals must sacrifice their freedom to the law and dominion of the state in order to receive it back again in a more perfect form. Such a state would thus promote freedom by way of order and order by way of freedom. True freedom would not be mere chance or chaos, but openness, creativity, and passion. Convinced that the facts of history testify to the truth of his dialectical view, Hegel boasts that his own Prussian society is, at last, the long hoped for fulfillment of this dialectical process.[8]

Notwithstanding incidental differences, Whitehead's view of history follows the same dialectical logic. For him, the overturn of pagan religion by the rise of Christianity served, alongside the Visigoth sack of Rome, as the quintessential destabilizing factor. Hence Whitehead frequently referred to what Hegel termed the negative as "Christians and barbarians." Christianity upset the pagan *status quo*, forcing dramatic changes that would eventually lead to an improved civil order. Sometimes, such changes have hinged upon the special ingenuity of a single individual. Hegel thus identified Julius Caesar as the executor of the negative in his conquest of Rome, and Whitehead's "Christians and barbarians" were often those genius scientists and inventors who, by challenging our accepted notions, propelled Western society toward new and exciting developments.

Not every evolutionary approach is dialectical, but they all tend to have their own ways of securing the openness of the future in relation to the relatively closed system of evolutionary progress. Celia Deane-Drummond, finding the process or Hegelian approach too abstract and mechanistic, argues instead for viewing evolution in terms of the theodramatics of Hans Urs von Balthasar (1905–88). As she sees it, many attempts to understand the story of evolution reduce it to mere *narrative*—a chain of events that follow one after another as by necessity. For a fuller and truer picture, we must view evolution—including humanity's role within it—in terms of *drama* as a dynamic, fluid, and open interweaving of necessity and contingency, objective and subject elements, natural events and free, unpredictable actors. "In other words," she explains, "the remarkable features of evolution are not so much the outcome of a grand design as an improvised script of a grand drama. This makes evolutionary history and that of humanity no less awesome."[9]

---

8 Hegel, *Lectures on the Philosophy of History*, 17.
9 Deane-Drummond, *Christ and Evolution*, 282; "Believing Deeply in Creation: Christ and Evolution as Theodrama," in *Evolutionstheorie Und Schöpfungsglaube: Neue*

Deane-Drummond's particular emphasis on the openness of the future connects to her call for action in regards, for example, to caring for the environment and our relationship with other creatures.

MORAL AND CULTURAL HORIZONS

The interplay of openness and closure is important when considering the moral and cultural horizons of evolutionary theology. In many cases, morality and/or culture serve as a means of opening up what could otherwise become an overly deterministic view of history.

Take for example the selfish gene approach mentioned in chapter 5. By leaning heavily on the notion that selfishness is embedded within the biological process of evolution, such an approach can appear deterministic. Richard Dawkins's solution to this problem leans on culture as the site of freedom. He insists that human culture is able not only to define itself but even to resist the biological imperatives of our genes.[10] Note the dialectical character of this move: if biological nature is through and though determined by violence, selfishness, and sin, then the only way that evolution can lead us forward is precisely by entering a new stage that counteracts or overcomes the past. Culture is the negative element that destabilizes the givenness of biology.

In contrast, many Catholic evolutionary theologians problematize this nature–culture dialectic.[11] After all, sociobiology itself sees culture as a product of our evolutionary development. One can see much of human culture as an extension of herd behaviors and in-group selfishness. More than that, culture has been and continues to be shaped by violence and competition. A mere cursory glance at the prevalence of English language, music, and literature throughout the world reveals the (often violent) imposition of British and American imperialism. Like the Greeks and Romans of old, we have spread our culture—along with many of its intellectual assumptions—as an extension

---

Perspektiven Der Debatte, ed. Hubert Philipp Weber and Rudolf Langthaler (Vienna: V and R unipress, 2013), 195. Her reading of Balthasar is heavily influenced by Ben Quash, *Theology and the Drama of History* (Cambridge: Cambridge University Press, 2005).

10 Dawkins, *Selfish Gene*, xiv, 201, 271.

11 Domning, *Original Selfishness*, 178; *Evolution*, 122, 125; Denis Edwards, *The God of Evolution: A Trinitarian Theology* (New York: Paulist, 1999), 68–70. See van den Toren, "Original Sin," 177; Stephen Duffy, "Genes, Original Sin and the Human Proclivity to Evil," *Horizons* 32, no. 2 (2005): 228.

and expression of power. Culture continues to evolve, and American culture has itself become more and more influenced by foreign elements, but such cultural evolution often imitates the logic of natural selection and thus makes progress by way of competition and the elimination of less successful cultural forms. Given that culture is itself an expression of the evolutionary struggle, how can culture ever overcome evolution's negative draw? This is not to say that culture is through and through corrupt. Yet culture alone cannot save us from our biological inclinations. In fact, culture can even enshrine and enforce evil inclinations in the manner of social sin.

Consequently, positive cultural development requires authentic moral development and vice versa. We have to learn to act according to positive impulses and good, rational goals so that our actions will contribute positively to the culture around us. At the same time, because one's individual freedom is never altogether separated from the broader cultural matrix, the promotion and development of good cultural values contributes to the moral actions of real individuals.

Utilizing the work of Sharon Welch, Cynthia Crysdale and Neil Ormerod argue that a theological understanding of human evolution points us toward what they term an "ethic of risk" rather than an "ethic of control."[12] That is, our decision-making must come to terms with the unpredictability and statistical nature of our evolving world. Rather than thinking that we can enforce or ensure success through the exercise of power, we must undertake moral action with the goal of opening up new future possibilities. In other words, for Crysdale and Ormerod, ethical action plays an integral role in securing the openness of the future. Right action takes the form of a daring undertaking, risking overwhelming odds, the entrenchment of social sin, and the real possibility of failure. Moral deliberation involves three kinds of risk. There is risk involved in the difficult labor of establishing the facts of the matter along with potential choices and outcomes. There is the objective risk of moral decision, which runs up against our own limitations; our individual actions can only ever partially combat the problem of evil. Finally, as moral agents we must weigh the heavy risk of taking responsibility for our actions and their consequences—whether intended or unintended, foreseen or unforeseen, good or evil.

In such a view, one's approach to ethics has significant consequences for cultural, social, and political structures. Attempting to "fix" the world through

---

12 Crysdale and Ormerod, *Creator God, Evolving World*, 110–16.

an "ethic of control," according Crysdale and Ormerod, tends to backfire. They point for example to the overuse of antibiotics and pesticides and the resultant evolution of resistant organisms.[13] In a sense, such an ethics wastes good intentions by failing to establish the promised stability. With its propensity for technocratic thinking, such an ethics can serve instead to enshrine problematic power dynamics. In contrast, an "ethic of risk" seeks by its very subtlety to build "new schemes of recurrence—regularities in which people can expect income, work, health care, and community on a stable basis."[14] In other words, it promotes societal structures that contribute positively to moral values without pretending to conclusively solve the world's problems or displacing the role of the individual as moral agent.

These points, while insightful, are not altogether novel. Yet Crysdale and Ormerod weave them together into a fabric of evolutionary theology in a way that skillfully places accent on the openness of the future. In fact, they argue that a Christian theological perspective further completes the "dialectic of chance and necessity, risk and control" by shifting us toward recognizing that everything we have, including all of our possibilities, are essentially gifts.[15] This leads to an "ethic of gratitude," an acknowledgment of the task of moral agency as openness to God's providential will. Ultimately, effective action for the future is that which cooperates with the grace of a loving and personal God.

In truth, from a thoroughly Christian perspective, both culture and morality must find their completion in divine grace. Neither element is free from the burdens of our evolutionary situation, and so we cannot look to either for a definitive solution to present-day problems. Both culture and morality open up to future possibilities, but without the addition of divine grace, even those possibilities may be restricted to a mere reworking of pre-given elements. The definitive openness of the future is nothing other than God's providential care for the world, which calls it to realize its own inner potential in ways that defy the limitations of our finite existence and the baggage of evolutionary selfishness.

Yet the theology of grace is an area where evolutionary theology still needs to grow. The powerless God of process theology and the self-limiting God of John Hick have tended to stymie the development of evolutionary theologies

---

13 Crysdale and Ormerod, 117.
14 Crysdale and Ormerod, 120.
15 Crysdale and Ormerod, 121–22.

of grace since they set strict limits on divine intervention. If, for example, grace is thought of only as an aspect of God's evolutionary creative activity, such an approach blurs the distinction between nature and grace. Patricia Williams, for example, suggests that the natural existence of altruism demonstrates that God's grace is intrinsic to the natural world.[16] Jack Mahoney takes a similar though more careful approach, arguing that human altruism derives from our creation in the image of God.[17]

Roger Haight provides a far more refined collapse of nature and grace. Haight reads the Bible, Augustine, Aquinas, and Rahner through the lens of Friedrich Schleiermacher, arguing that grace is best understood as the experience of God's creative presence. This is not so much a personal relationship with God but rather a feeling of being dependent on a transcendent other that is at least *not less than* personal. We interpret this subtle and passive presence of God in our consciousness as God actively revealing God's self. Yet this means that we have a tendency toward anthropomorphically "projecting God as a person interacting with human subjects."[18] In reality, according to Haight, all revelation and grace derive from this experience of dependence, not from any active divine intervention in the world. Grace is essentially nothing other than a natural, universal, and quiescent experience of God's loving presence as Creator. This loving presence is experienced in "four key words: *acceptance, forgiveness, healing, and energizing*," but Haight does not intend any of these to imply a supernatural order over and above the natural order of creation.[19] God does not forgive us for particular sins, for example. Rather, God's presence itself is felt as a general sense of forgiveness that heals our feelings of guilt.

In this short space we can neither detail a complete critique of such a collapse of nature and grace nor enter into the complex question of divine intervention. Suffice it to say that Haight departs significantly from Karl Rahner despite holding him as an influence. While Rahner does argue against viewing the creation of the human soul as a miraculous event wherein God altogether disrupts the natural order of the world, he never discounts divine

---

16 Williams, *Doing Without*, 155.
17 Mahoney, *Christianity in Evolution*, 44.
18 Roger Haight, *Faith and Evolution: A Grace-Filled Naturalism* (Maryknoll, NY: Orbis, 2019), 106–108.
19 Haight, 108–109.

intervention as such.[20] For Rahner, God's interaction with the world is not simply reducible either to some general existential feeling of dependence or else to God's mere presence as Creator or Prime Mover. Most importantly, while Rahner affirms the gratuitous ubiquity of divine grace—grace is at least present as an offer to each and every individual—he is careful to maintain the distinction between nature and the supernatural.[21] In Rahner's view, if grace is nothing more than the experience of God as Creator, then grace loses its real gratuity and significance.

In short, there is a vital difference between the axiom that grace perfects (or completes) nature[22] and the claim that grace *is* nature, an aspect of nature, or something guaranteed and given by necessity alongside nature. This is the case not only because Christianity understands grace in terms of a personal relationship with the Triune God but also because the transcendence of grace is what grounds its ability to save us from impossible situations. If grace were nothing more than a part of nature, on what basis could it rescue us from our own evolutionary selfishness?

In fact, it is grace that reveals the fundamental goodness of our nature by completing it. Since grace elevates nature, it shows that nature is not reducible to mere selfishness. Rather, as Mahoney means to show, there are real, good, and even altruistic tendencies to our nature, and grace takes these up, energizes, and reconfigures them according to the model of Jesus Christ. In this way, even a tribalistic in-group altruism can become the basis of a self-sacrificing Christian charity that knows no boundaries or exclusions.

Culture and morality, therefore, must be completed by divine grace precisely so that by means of these we can transcend the impediments of nature and concupiscence. As Gabriel Daly and Denis Edwards point out, our problematically self-interested biological instincts and impulses only become sinful when we do not allow them to be healed by divine grace. Culture too is implicated in this need for grace. While sin can take the form of allowing selfish biological urges to dominate our better cultural inclinations, it can also take the form of allowing distorted cultural

---

20 Rahner, *Hominisation*, 66, 100–101.
21 Karl Rahner, "Concerning the Relationship between Nature and Grace," in *Theological Investigations*, trans. Cornelius Ernst, vol. 1 (Baltimore: Helicon, 1961), 307–308.
22 Thomas Aquinas, *Summa theologiae* I, q. 1, a. 8, ad 2.

imperatives to overwhelm our healthy biological impulses (think socially generated eating disorders).[23] For Daly, this means that the history built up by human culture is at one and the same time fallen history and the history of salvation:

> Where we or our ancestors have failed to allow God's grace to heal us, we construct historical monuments to our sin, neatly and grimly symbolised in the fact that we build concentration camps. Where we or our ancestors have responded to grace, we have erected monuments to our graced nature such as hospitals.[24]

In this way, culture is neither innocent nor inherently corrupt. It is a site of open possibilities, but those possibilities must be realized by the reception of grace.

TECHNOLOGICAL HORIZONS

If evolutionary theology directs us onward, it is only logical that technology should constitute a key horizon of development. After all, given the phenomenon of globalization, human biological evolution has taken an altogether different shape. It is not that we have ceased to evolve. Yet Darwinian natural selection cannot operate in full on a population where even those born infirm are often able, thanks to cultural, technological, and especially medical developments, to live, to thrive, and to reproduce. Many evolutionary theologians thus look beyond biology for humanity's future, and some in particular focus on technology as the most promising site of rapid and consequential development.

Technological optimism flourished in the late nineteenth to mid-twentieth century. In 1932, Elie Foure boldly proclaimed, "It is technology that will save the spirit, because technology is, to date, the last incarnation of the spirit."[25] In a similar way, Teilhard de Chardin saw technology as a vital tool for spiritual development. He marveled at the "scientific triumph" of the atomic bomb and at the prospect of gene editing, and he argued that the mechanization

---

23 Edwards, *God of Evolution*, 68–69.
24 Daly, *Creation*, 144–45.
25 Quoted in Pierre-Louis Mathieu, *La pensée politique at economique de Teilhard de Chardin* (Paris: Éditions du Seuil, 1969), 78 n. 65.

of production would free people for higher mental activities.[26] Technological development was in essence a *spiritual* task—a symptom of the noosphere's ongoing evolution.

Teilhard's particular confidence about the future would face new challenges today. Technology's rapid expansion has made it all the more difficult to predict any particular outcomes. Technology has thus become a symbol of unlimited progress toward an *open*, unfettered future; a forward movement that lacks any clear, determinate end.[27] Some theorists view these open possibilities with wonder. For example, if technology can continue to delay death, can it make us effectively immortal? Others, in contrast, see technology as a horizon of unforeseen dangers. Fueled by postmodernity's general disillusionment with the cult of progress, many today look upon technology with a growing degree of suspicion. Not only have promises of flying cars and cities on the moon failed to come true, but there are real ways that technology can pose serious moral, political, cultural, and even spiritual challenges. Celia Deane-Drummond points for example to how, in the late 1950s, gynecologists disregarded medical risks and used abusive marketing tactics to convince women that they needed to take estrogen in order to rescue their own womanhood from the process of menopause.[28]

Evolutionary theologians therefore face the difficulty of expressing an integral view of technology that takes seriously its positive contributions without making technology a replacement for authentic spiritual development. How can we advocate for new medical technologies without reducing the Christian quest for eternal life to a mere endless technological postponement of death? In particular, Deane-Drummond recommends a robust and revitalized Christian eschatology centered upon the resurrection and finding perfection in God as an antidote to a merely secular and unregulated quest for technological self-perfection.[29] After all, approaching technology's challenges today calls for more than dealing with particular moral issues on a case by case basis. It requires a wholistic understanding of technology and the human person as such.

---

26 Pierre Teilhard de Chardin, *The Future of Man* (New York: Harper and Row, 1964), 144–46; 171–72; *Phenomenon of Man*, 250; *Activation*, 170.

27 Peter J. Bowler, *Progress Unchained: Ideas of Evolution, Human History and the Future* (Cambridge, UK: Cambridge University Press, 2021), 254.

28 Celia Deane-Drummond, "Future Perfect? God, the Transhuman Future and the Quest for Immortality," in *Future Perfect?: God, Medicine and Human Identity*, ed. Celia Deane-Drummond and Peter Scott (London: T and T Clark International, 2006), 170–71.

29 Deane-Drummond, "Future Perfect," 182.

In contrast, coming from a process theology perspective, Philip Hefner approaches this situation from the angle of humans as "created co-creators." In his view, technology is a mirror of humanity's innate creativity and imagination.[30] Yet Hefner also sees such imagination as the core of spiritual experience, so much so that he sees religion itself as a form of technology.[31] The point is that religious rituals are part of how we cocreate meaning in our lives and transcend the limits of the actual by imagining the perfections of eschatological existence here and now. This does not mean that unfettered technological development is inherently good, but it does mean that such development implicitly carries religious significance and impacts us on a much deeper level than we might expect. Hefner goes so far as to say that "when we participate in this drive for new possibilities, we participate also in God," and he therefore avers that the existential restlessness of technological imagination is "a means of grace."[32]

Hefner's extremely positive evaluation of technology resonates with *transhumanism*, though he does not don this label himself. Coined by Julian Huxley in 1957, this term has come to signify an ideology of technological progress. Archimedes C. Articulo describes it as "the belief that humans must wrest their biological destiny from evolution's blind process of random variation and adaptation and move to the next stage as a species, favoring the use of science and technology, especially neurotechnology, biotechnology, and nanotechnology, to overcome human biological limitations."[33] Two influential contemporary transhumanists are Ray Kurzweil and Nick Bostrom.

Concerned with its almost uncritical appreciation of technology, evolutionary theologians often use transhumanism as a foil. Ilia Delio, for example, maintains something of Teilhard's optimism but entertains suspicions that transhumanism leads toward an obsession with individual self-modification rather than genuine communal revitalization. Because of this, she prefers the more Teilhardian label of *ultrahumanism*.[34] As she sees it, technology

---

30 Hefner, *Technology*, 44–45.
31 Hefner, 46–48.
32 Hefner, 84, 86.
33 Archimedes C. Articulo, "Towards an Ethics of Technology: Re-Exploring Teilhard de Chardin's Theory of Technology and Evolution," *Open Journal of Philosophy* 4, no. 4 (November 3, 2014): 519. Julian Huxley, "Transhumanism," in *New Bottles for New Wine* (London: Chatto and Windus, 1957), 13–17.
34 Ilia Delio, "Transhumanism or Ultrahumanism?: Teilhard de Chardin on Technology, Religion and Evolution," *Theology and Science* 10, no. 2 (2012): 153–166.

contributes positively to evolutionary progress when it brings people together while maintaining the dignity of the human person and the natural world.

Despite significant differences, perhaps what unites many evolutionary theologians is the recognition that technology belongs to humanity in such a deep way that it cannot simply be set aside as something alien. In fact, technology is in its essence an expression of who we are. As Karl Rahner points out, we are the being whose own development has been handed over to itself.[35] We have been constructing ourselves for as long as we have been free, because the history of our freedom has always had consequences for humanity's development. The patterns of human migration, friendship, and conquest have shaped our very genetics. Likewise, the tools we make, the foods we cultivate, the animals we domesticate—all of these and more are part of our technological heritage. While our means of manipulation have developed dramatically in recent decades, the fact that we are technological manipulators has remained the same since the dawn of humanity. After all, genetic manipulation began as far back as cultivation, as many of the crops we eat have been carefully selected over millennia to benefit the needs of the human race.

We must keep this in mind as we consider today's most daring technological innovations. We already have the technology to permanently modify specific, targeted portions of the DNA of living individuals. CRISPR-based gene editing technologies continue to develop at an astounding rate.[36] Gene editing carries amazing potential. In many cases, it can literally bring sight to the blind and cure congenital diseases. There are of course significant moral considerations to be made—even more so when considering the manipulation or abortion of embryos and the editing of inheritable DNA (known as *germline modification*). Yet the moral issues involved with particular applications do not legitimate a blithe dismissal of such technologies in general as being against nature or against God's creation. The truth is that the nature that God created is essentially open to modification. A realistic and critical understanding of technology must go far beyond any dualism of technology

---

35 Karl Rahner, "The Problem of Genetic Manipulation," in *Theological Investigations*, trans. Graham Harrison, vol. 9 (New York: Herder and Herder, 1972), 227.

36 CRISPR is short for "clustered regularly interspaced short palindromic repeats." Certain bacteria utilize gene-editing molecules to insert enemy DNA into these sequences in order to more quickly recognize threats. Scientists are now able to use modified forms of these molecules to arbitrarily insert specific DNA sequences into a chromosome.

versus nature. If technology is an expression of ourselves, then who is the self whom we express by means of such new technologies?

In other words, technology's potential for good or for evil is a refraction of our own inner potential as spiritual beings. This is not to say that technology is neutral. Rather, inasmuch as everything we do has real moral consequences, so also is our technological self-expression marked by both good and evil. For example, economic injustice and the abuse of laborers did not begin with the Industrial Revolution and its mechanization of production. Rather, this technological revolution optimized and systematized existing forms of injustice like never before. Even as the United States gradually outlawed the practice of slavery, the basic concupiscent evils that birthed this institution remained ready to take new form in the industrial evolution of capitalist society. In this sense, technology does not produce new evils. Rather, when technology becomes isolated from genuine moral deliberation, it recycles old evils with great efficiency into new and ever more destructive manifestations.

As a consequence, we must now more than ever face the technological horizons of human evolution with a sober theological consciousness of its moral and anthropological implications. The question is no longer, "Can a new technology rescue us from inequality, injustice, or suffering?" Technology cannot save us from evil any more than it can save us from ourselves, for in it we face our very own evil—the reflection of original sin. Rather, the question becomes, "How can advances in technology be shaped by human and ethical concerns? How can we identify positive and negative aspects of technology's influence on our lives and promote or mitigate these factors so as to make use of technology in the most responsible manner?" In other words, the question of technology is another form of the question of humanity itself. "Who am I, and what am I forming myself to become?" modulates into "How do I express myself technologically, and how does this technological self-expression shape my own coming-to-be?"

As Ronald Cole-Turner points out, genuine Christian hope is more costly than the technological future sought by secular transhumanism.[37] We must

---

37 Ronald Cole-Turner, "Religion, Technology, and the 'Future' of Evolution," in *Evolutionstheorie Und Schöpfungsglaube: Neue Perspektiven Der Debatte*, ed. Hubert Philipp Weber and Rudolf Langthaler (Vienna: V and R unipress, 2013), 299; "Transhumanism and Christianity," in *Transhumanism and Transcendence: Christian Hope in an Age of Technological Enhancement*, ed. Ronald Cole-Turner (Washington, DC: Georgetown University Press, 2011), 199.

be careful not to let our technological aspirations displace Christian eschatology with its moral and spiritual demands, demands that place a central emphasis on self-gift, conversion, and transformation rather than on mere physical, biological enhancement.

ECOLOGICAL HORIZONS

The most pressing issues of our day make it all the more vital to coordinate our technological solutions with deep moral deliberation guided by Christian eschatological hope. In particular, among present-day crises the rapidly accelerating situation of global ecological devastation stands in a category of its own. It is in many ways a distinctive trademark of our postmodern era. Nevertheless, it involves such an array of diverse factors and causes that it is perhaps better understood as a broad collection of interrelated crises. Climate change feeds into mass extinctions, population collapses, and the loss of biological diversity, but these have other causes as well. They are linked also to deforestation, contamination, industrialized farming, the diaspora of invasive species, and a broad array of technological, economic, and social factors. These factors diversely affect the ecology not only of uncultivated lands but also of cultivated and urban environments. Even humanity's habitats are integral to the ecology of our world. Thus, contemporary ecological science must consider a host of challenges ranging from forest management to sustainable city planning and from wildlife protection to safeguarding public health.

Today's ecological challenges also have a significant impact on present-day experiences and concepts. These challenges both stem from and feed back into our attitudes about our natural environment. The rising demand for organic and non-GMO foods, for example, exemplifies a growing sense of alienation. Many feel their experiences are defined by technology *in contrast* to an ideal of nature, and this leads them to reject technology in favor of such an ideal. As a result, many understand the natural world in terms of a conflict between nature and technology. This situation poses a wealth of difficulties not only in terms of our identity and the theological relation between God and the world but also in terms of environmentally responsible religious praxis.

A growing wave of theologians are responding to today's ecological challenges. Their writings call attention to problematic, environmentally destructive attitudes and themes within theology and seek to highlight, rediscover, and develop themes and attitudes that place the appreciation and protection of the natural world at the center of theology and Christian praxis. Many proponents of ecological theology label their work *ecotheology*.[38] For the sake of simplicity, we will use this term to speak broadly of ecological forms of Christian theology in general. Keep in mind, however, that not every ecological theologian would accept this term, and there are important differences among ecological theologians that cannot receive adequate treatment here.

Importantly, not every ecotheology is also evolutionary. Nevertheless, many evolutionary theologians dabble in ecotheology, and many ecotheologians also make important appeals to evolution. This connection is far from incidental. As we have seen, the close interrelation between humans and other creatures represents a central conviction of the evolutionary paradigm. Evolutionism reinterprets humans in view of their place within the natural world; evolutionary theology typically also reinterprets the natural world in light of this revised understanding of the human race. As our very consciousness of our place within this world continues to evolve, it provides new insights for appreciating the natural world's theological significance and reconfiguring its relationship with humanity both practically and conceptually. Because of their substantial overlap, much to be said concerning evolutionary ecotheology applies first and foremost to ecotheology broadly speaking. Without attempting to fit anyone into strict categories, our discussion here will take a broad approach. The ecological theologians we will consider all make significant use of evolution, but not all of them might ordinarily be considered evolutionary theologians.

Ecotheological Diagnoses of the Problem

Ecotheologians often begin with a diagnosis that sees our present-day ecological threats as symptoms of an underlying *theo*logical problem. Such a diagnosis makes this discourse more than environmentalism in a religious wrapping. By arguing that Christian theology has contributed substantially

---

38 For an overview of the history and major themes of ecotheology, see Dieter T. Hessel and Rosemary Radford Ruether, eds., *Christianity and Ecology: Seeking the Well-being of Earth and Humans* (Cambridge, MA: Harvard University Press, 2000), xxxiii–xlvii. See also Celia Deane-Drummond, *A Primer in Ecotheology: Theology for a Fragile Earth* (Eugene, OR: Cascade Books, 2017).

to our ecological problems, ecotheology inaugurates its quest to develop a robust theological response to counteract this baleful heritage.

Unfortunately, in making such a diagnosis, ecotheologians have often relied on Lynn White Jr.'s deeply flawed and unscholarly assertions.[39] In a famous 1967 essay, White accuses medieval European Christians of inaugurating a violent history of antagonism between humanity and the natural world. He points in particular to their use of technologically advanced farming equipment.[40] Ignoring a wealth of contradictory evidence, White construes this development as flowing naturally out of the biblical command to subdue the earth and have dominion over all living things (Gen 1:28). Ecotheologians vary in the degree to which they accept or challenge White's assertions. Nevertheless, these claims have hit hard and left their mark despite their lack of any sound historical basis.

In reality, although medieval Christian attitudes and actions toward the natural world were not blameless, neither were they resoundingly negative or adversarial.[41] The Bible and Christian tradition carry positive as well as negative assertions about the natural world. Moreover, medieval attitudes and actions were shaped by a range of influences such as classical ideas, class structure, feudal divisions of land, and the ever-present threat of famine and disease. Although medieval Christianity did not instill a strong sense responsibility toward the natural world as such, it did carry a sense of responsibility *to* God *for* the world. A deeply hierarchical viewpoint served to order all matters, activities, and resources toward a higher purpose.[42] In this way, humanity's right to *use* nature was not understood as a right to *abuse* it.

Many ecotheologians instead highlight the destructive influence of later, postmedieval developments. Without denying the climatic impact of earlier human activity,[43] they point to the exponential growth of humanity's exploitation of the natural world since the sixteenth century. Earlier forms

---

39 See for example John B. Cobb Jr., *Is It Too Late?: A Theology of Ecology* (Denton, TX: Environmental Ethics Books, 1995), 31–33.

40 Lynn White, "The Historical Roots of Our Ecologic Crisis," *Science* 155, no. 3767 (1967): 1205.

41 See for example John Aberth, *An Environmental History of the Middle Ages: The Crucible of Nature* (London: Routledge, 2013), 8.

42 Bonifazi, *Soul of the World*, ii.

43 Studies have argued that humans have contributed to climatic changes for thousands of years. See for example Jed O. Kaplan et al., "Holocene Carbon Emissions as a Result of Anthropogenic Land Cover Change," *The Holocene* 21, no. 5 (August 1, 2011): 775–791.

of exploitation are dwarfed by the vast ecological impact of colonization, urbanization, globalization, and especially industrialization. In terms of human values, therefore, the present-day crises owe less the heyday of European Christianity and more to its general decline. "As the demands of Christian discipline relinquished their hold upon the West," Conrad Bonifazi explains, "the idea of the guardianship of the earth came to be seen as an obstacle to economic growth."[44] Planetary exploitation expanded exponentially alongside the major political, philosophical, and economic shifts that heralded the Renaissance and ushered in the modern era. In particular, the rise of capitalism through the conquest and violent development of foreign lands propelled ideals of private ownership, personal prosperity, and the mechanization of production.[45] In this way, the ecological crisis is also very much a social crisis; it goes hand in hand with the ever-expanding gap between rich and poor. The concentration of wealth and power fuels and is fueled by a system whereby the few enjoy the unchecked exploitation of the earth's resources and leave the weaker majority to suffer the greater portion of the ecological consequences.[46] To the extent to which society calculates the value of human persons in terms of their usefulness, it should come as no surprise that nature too "becomes a mere object with no meaning or vocation beyond its utilitarian value to human beings."[47]

Despite its limitations, White's thesis continues to influence ecotheology in an especially important way. His indictment of Christianity popularized the view that present-day ecological problems are deeply rooted in traditional religious ideas, concepts, and language. Ecotheologians such as H. Paul

---

44 Bonifazi, *Soul of the World*, iii. Pernoud notes for example that only at the end of the fifteenth century did Philip the Handsome codify "the *jus utendi et abutendi*, the right of use and abuse, in complete contradiction to the customary law of that time." Régine Pernoud, *Those Terrible Middle Ages: Debunking the Myths*, trans. Anne Englund Nash (San Francisco: Ignatius, 2000), 100.

45 Michael Northcott explores these developments in detail in *The Environment and Christian Ethics* (Cambridge: Cambridge University Press, 1996), 40–85. Though Anglican himself, Northcott sees the rise of Protestantism as partly responsible for Christianity's negative contribution to these developments. See also Daniel Cowdin, "The Moral Status of Otherkind in Christian Ethics," in *Christianity and Ecology: Seeking the Well-Being of Earth and Humans*, ed. Dieter T. Hessel and Rosemary Radford Ruether (Cambridge, MA: Harvard University Press, 2000), 265.

46 Rosemary Radford Ruether, *To Change the World: Christology and Cultural Criticism* (New York: Crossroad, 1981), 58–59.

47 Sallie McFague, *Super, Natural Christians: How We Should Love Nature* (Minneapolis, MN: Fortress Press, 1997), 59.

Santmire see the ecological crisis as at least in large part *"a crisis in values."*[48] Our attitudes and practices in regard to the natural world are deeply shaped by our cosmological and theological assumptions, which are themselves shaped by the concepts and language that we use in order to express our deepest commitments. In view of this, ecotheologians argue that theology can and must become an essential part of the solution.[49] If Christian thought can promote ecologically destructive habits, then it is capable also of promoting the opposite. If problematic ideas, concepts, and phrases have misframed humanity's relationship with the natural world in terms of conflict, then Christianity must recover a sense of communion, a positive meaning of relationship that lies at the heart of creation itself.

Ecotheological Appeals to Evolution

To this end, evolution often plays important conceptual roles. In fact, like evolutionary theology in general, evolutionary ecotheologies often follow to varying degrees the patterns set by Whitehead and Teilhard de Chardin. The extreme optimism of the latter figure does not always mesh well with environmentalists' dire predictions of the future, but his views do provide a strong sense of creation's unity and organic interrelation.[50] Some distinctly Teilhardian ecotheologians include Conrad Bonifazi, H. Paul Santmire, David Toolan, and Denis Edwards. Yet process theology has perhaps held a broader influence on ecotheology, especially through the works of John Cobb Jr. and Charles Birch.

Many ecotheologians from both of these angles appeal to evolution in particular because of the way it forces us to rethink the relationship between humanity and the natural world. As we have seen, the discovery of natural selection destabilized the old anthropocentrism, making humanity's similarity to rather than difference from other animals the key starting point for understanding our place in the universe. Even though this led to the rise of new forms of anthropocentrism, it also contributed to the recognition that

---

48 H. Paul Santmire, *Nature Reborn: The Ecological and Cosmic Promise of Christian Theology* (Minneapolis, MN: Fortress Press, 2000), 3.

49 Willis Jenkins, "After Lynn White: Religious Ethics and Environmental Problems," *The Journal of Religious Ethics* 37, no. 2 (2009): 285–86.

50 See for example Robert Hale, *Christ and the Universe: Teilhard de Chardin and the Cosmos* (Chicago: Franciscan Herald, 1973), vi.

the natural world has its own intrinsic meaning and value apart from human use and consumption.

In view of this, many ecotheologians interpret the scientific theory of evolution as revealing the fundamental community of all creation. If humans are no longer the exception but rather a part of the same evolutionary paradigm, then this points to a deep sense of relationality that must displace our prideful self-aggrandizement. Some theologians express this sense of community in a relatively Teilhardian manner, with its tendency for strong teleology and revised anthropocentrism. Thus, the influential Reformed theologian Jürgen Moltmann sees all creatures as united not only in their origin but also in their teleological orientation. Humans are both members of this community and privileged mediators between God and creation.[51] Yet many other theologians look to process theology for its more extreme decentering of the human subject and/or its absorption of the God–world relationship into the process of becoming. As we saw, Whitehead extended ordinarily subjective terms such as "feeling" and "urge" to all actual occasions—even rocks and atoms. Following in this vein, Sallie McFague argues that all existing things exhibit some degree of subjectivity and that this means that all things exist only through interdependence with all others.[52] For McFague, Christians are called thus to extend Jesus's countercultural love to *all* subjects, loving all creatures in a way analogous to loving other humans. This leads to "a conservation ethic of sustainability—an ethic of preserving and sharing the earth's bounty for all its creatures—[which] is implicit in Genesis."[53]

From a strictly scientific standpoint, however, such appeals to evolution are rather opportunistic. Because of this, Lisa H. Sideris argues that many ecotheologians build their arguments on problematic, one-sided, and myopic appeals to science. Their frequent emphasis on the community of all creation, for example, leads them to value scientific points only to the extent to which they portray nature in a positive, holistic, and communitarian light. They assume that nature inherently tends toward harmonious relationships and that chaos, disruption, and catastrophic death stem primarily from human tampering. Rosemary Radford Ruether, for example, defines evolution not as

---

51 Jürgen Moltmann, *God in Creation: A New Theology of Creation and the Spirit of God*, trans. Margaret Kohl (San Francisco: Harper and Row, 1985), 190.

52 McFague, *Super, Natural Christians*, 2–3; *The Body of God: An Ecological Theology* (Minneapolis, MN: Fortress Press, 1993), 76.

53 McFague, *Super, Natural Christians*, 41, 164–67.

a bitter competition for survival but rather "as a process *aimed* at community existence and cooperation."[54] For Sallie McFague, "The evolutionary perspective is intrinsically protoaltruistic, displaying features such as give-and-take, sharing, symbiosis, and life and death."[55] As Sideris points out, such claims derive more from Romanticism than from Darwin.[56]

Sideris credits such anemic appeals to evolutionary science with contributing to self-contradictory and impractical approaches to environmental ethics. In glossing over the violent and competitive elements of evolution, they sidestep many important and difficult questions. This leads for example to a tendency to rely upon anthropocentric judgments for determining the moral value of particular environments, species, populations, or individuals. This reliance remains largely implicit in the work of Sallie McFague. She sees sentience and intelligence as constituting a higher value and, because of this, she allows greater significance for conscious beings than for unconscious, unintelligent, or inanimate beings. Given such a premise, bears must still have a higher value than jellyfish. For Sideris, this means that McFague's talk about loving all things has become lost along the way: "An ethic of radical relationality is abandoned in favor of one that assumes different 'levels' of being, and yet the justification for different levels of being is never provided."[57]

In contrast, process ecotheology makes such a hierarchy explicit but relative.[58] On the one hand, in order to relativize the significance of human persons, John Cobb Jr. goes so far as to deny that human life is sacred.[59] On the other hand, he allows humanity to claim a higher value to the extent to which we bear a greater capacity for the enjoyment of feeling—i.e., for experiencing the beauty and variety of the world process.[60] Humans are not valuable absolutely but only relatively. To put it another way, our value is mediated to us as a byproduct of the intrinsically valuable process of evolutionary development.

---

54   Lisa H. Sideris, *Environmental Ethics, Ecological Theology, and Natural Selection* (New York: Columbia University Press, 2003), 47–49.

55   Sallie McFague, *A New Climate for Christology: Kenosis, Climate Change, and Befriending Nature* (Minneapolis, MN: Fortress Press, 2021), xi.

56   Sideris, *Environmental Ethics*, 69–70.

57   Sideris, 79.

58   Sideris, 116.

59   Cobb, *Is It Too Late?*, 55–56.

60   John B. Cobb Jr. and David Ray Griffin, *Process Theology: An Introductory Exposition* (Philadelphia: Westminster Press, 1976), 63.

For Sideris, such ecological approaches can become lost in impracticable ideals because they attempt to apply one, singular, consistent ethic for both human relationships and the natural world.[61] For example, McFague sees Christian love as inherently applicable to the environment because she extends subjectivity to all creatures. This leads to a glaring internal contradiction: however much a Christian loves lettuce, this will hardly prevent one from enjoying a salad. In contrast, Christian moral considerations in regard to persons are often inherently singular; one should respect an individual human *as* an individual, not merely as a member of the species or an organelle of human society. Yet environmental ethics often requires decisions to be made that place a categorically higher value upon species. Sideris points to an important example wherein ecologists killed fifteen thousand non-native but wild goats in order to prevent the extinction of three native species of plants.[62]

Sideris's differing approach highlights an important decision within ecotheology. Many evolutionary ecotheologians are influenced by or friendly toward process theology precisely because, following White's diagnosis, they perceive our ecological crises as being particularly rooted in theological or philosophical cosmology.[63] From such a standpoint, they see ecotheology's principal task as a matter of reframing our understanding of the natural world and its relation to God. Process theology, with its penchant for reconceptualizing God in terms of development and self-revelation, is especially apt for this kind of approach. Hence a wide array of ecotheologians see evolution primarily as providing justification for such a conceptual project.

Sideris illustrates a significantly different approach. Prioritizing the "land ethics" of Aldo Leopold's *A Sand County Almanac* (1948) as well as the insights of James Gustafson and Mary Midgley, Sideris does not root her claims in cosmology or the nature of God, nor does she propose a singular ethic for all relationships. "An appropriate environmental ethic turns on distinctions."[64] While eschewing overly anthropocentric registers of value, we must be willing to make decisions that, for example, prioritize species over individuals and native species over those which have been introduced. The point is that a love of nature is possible, but this must mean accepting nature and life-forms as gifts, which carry some intrinsic measure of value beyond their similarity to

---

61 See for example Cobb, *Is It Too Late?*, 49–50.
62 Sideris, *Environmental Ethics*, 180.
63 Jenkins, "After Lynn White," 291–92.
64 Sideris, *Environmental Ethics*, 267.

ourselves.[65] Such a love must accept nature for what it is without glossing over its violence and competition. The result should be a practical ecology with a balanced approach to intervention, recognizing that humans cannot solve all of nature's problems and in many cases should not try, lest undue intervention lead to greater problems through the disruption of natural processes.

We cannot in this space enter any more fully into the complex questions and issues of ecology. This cursory examination has shown, however, that the dialogue between evolutionary theology and ecology is far from complete. A deeper understanding of evolution not only contributes to a better appreciation of the natural world but can also drive our ethical and ecological actions toward more practical outcomes.

CONCLUSION

There is much more that we cannot cover here. We have not had the space to discuss Teilhard's vision of cosmic worship, John Haught's ideas about beauty, or Jerry Korsmeyer's preparations for extraterrestrial encounters, to name but a few examples. Nevertheless, this short sampling of future horizons has illuminated the vital function of the future within evolutionary theology.

As we said at the start of this chapter, a vision of the future ultimately allows many evolutionary theologians to issue a call to action for the present day. We have looked at a broad array of ideas about the future. For some theologians, the future is an open site of moral and/or cultural development; for others, it is a creative expanse of technological development; for still others, it is a vista of ecological harmony. Yet in each case, the future in some sense makes a demand upon us today, an imperative to act so as to make the future a present reality.

In the end, a strong evolutionary imperative requires a robust theology of grace. Without such a focus, the message of an evolutionary theologian can easily devolve into either a Pelagian insistence on hard work and effort or else a baseless trust in the mere upward trajectory of evolution itself. Grace, after all, constructs its own history of salvation, which builds upon but is not reducible to the history of the evolution of the world. In fact, it is precisely grace that undergirds the imperative character of theology's call to action,

---

65 Sideris, 252–53.

for the offer of grace presents itself as an absolute and fundamental decision: a Yes or No to God's loving demand for action. To heed the call is not a mere matter of passively evolving into some higher sort of humanity. Rather, grace comes to us precisely so that we can more truly become the God-beloved creatures that we already are. To be human is not merely a biological given, but—in this sense—a *theological* imperative.

Epilogue

An Invitation

We come at last not to a conclusion but a crossroad. There is much more to explore, and evolutionary theology is far from a closed book. It began with the simple premise that science and theology have something to share with one another. This developed into a rich and fertile field of theological insight. The full harvest has yet to be gathered.

Our explorations have opened up a broad array of evolutionary theological points of view. Although for the most part we have not been able to delve at any depth into critiques of these views, some important aspects of critique have presented themselves as a necessary consequence of delineating their key claims and commitments.

As a follow-up, I wish here to lay out three considerations of special importance for the future of evolutionary theology. First, I will argue that, far from helping, process theology has sidetracked evolutionary theology by its commitment to a problematic metaphysics. Second, the engagement between science and theology has really only just begun. The future of evolutionary theology requires a deeper, more attentive conversation between these two disciplines. Last, I will argue that the question of theodicy should be approached differently, placing emphasis not on justifying God but rather on understanding human evil and promoting healing, liberation, and pardon.

MOVING BEYOND PROCESS THEOLOGY

As we have seen, process theology has become more and more entwined with evolutionary theology in general. Yet the fact that evolutionary theology has existed without it shows that process theology is not, in the end, necessary. I hope that the foregoing analysis of Whitehead has shown that process theology is indeed fascinating and perhaps even beautiful but also deeply problematic.

In particular, a process approach to God fundamentally undermines the broader Christian tradition and makes God into nothing more than a cog in the system of creative advance or Hegelian dialectic. While its intention is to highlight God's intimate relationship to the world, it does so only by relegating divine transcendence to an abstract notion of God's primordial nature. God becomes transcendent merely *in principle* in order to be intimate *in fact*. But what this really means is that God's intimacy has no merit. Whitehead's God is near to us simply because God *has no choice in the matter*. In fact, the world is effectively fodder for the self-development of this deity. Or, alternatively, God is merely another way of looking at the world itself; God is the world seen from the angle of the coming-to-be of value. As Jacques Maritain notes, the God of Hegel or Whitehead is

> a God tied to the order of nature or to the evolution of the world; a God who is nothing more than the supreme guarantee and justification of that order or of that evolution; a God who is responsible for this world but without the power to redeem it, and whose inflexible will, that no prayer can reach, is pleased with and condones all the evil as all the good of the world . . . a God who blesses iniquity and slavery and misery, and who sacrifices man to the cosmos, and who makes of the tears of children and the agony of the innocent a mere ingredient of the sacred necessities of the eternal cycles or of evolution—with no after-life where His goodness mends the ravages made in His work by created freedom.[1]

Nietzsche proclaimed the death of Hegel's God—and for good reason. Any God that is merely a metaphysical necessity, a part of the machine of ontotheology, cannot be the living God of the Bible and tradition.

If Christianity is nothing more than a statement about God's intimate relationship with the world, then process theology is perfect. But if it is more than that, if it is a personal and dynamic relationship with God, if this God is not merely an exemplar of the evolutionary progress of the world but a

---

[1] Jacques Maritain, *Moral Philosophy: An Historical and Critical Survey of the Great Systems* (London: G. Bles, 1964), 191.

personal reality to whom we relate and who relates to us as Father, Son, and Holy Spirit, then process theology simply falls short. The problem is not that process theology goes too far, but that it does not go far enough. It establishes God's intimacy with the world precisely by abstracting it, making it no longer the bridegroom and the bride but at best the pilot and the copilot.

Proponents of process theology often make the process God seem attractive by misrepresenting traditional views. Process theology has much to critique about classical theism, but Christianity is *not* classical theism. Rather, Christianity is the intimate romance between God and the world in the incarnate Jesus Christ. While it is true that patristic and medieval theologians consider God to be utterly transcendent and beyond the world, they hold this in paradoxical tension with the most profound intimacy or immanence. In Augustine's famous phrasing, God is *interior intimo meo et superior summo meo*, "more inward than my inmost and higher than my highest."[2] It is precisely because God is transcendentally other than the world that God can freely enter into the world as the incarnate Jesus Christ without sacrificing anything of divinity. In this way, moreover, God fully understands our sufferings from the inside, and yet God stands above all suffering and is able to help us in our need. The God of the Bible is a God who constantly intervenes out of love, the saving God who almost cannot help but care for God's people.[3]

Moving forward, evolutionary theologians need to be more circumspect about their use of process thought. Rather than taking its claims at face value, one must recognize its relation to broader influences and controversies, seeing for example how its critique of divine transcendence applies to Aristotle and not to the God of Christian tradition. When process thought is utilized, theologians should be more upfront about its role within their thought and the extent to which their conclusions rely upon a process understanding of God. Only in this way can readers make an informed and clear decision as to whether the proposed benefits of the process approach outweigh its significant consequences for traditional Christian belief and practice.[4]

---

2 *Confessions* III, 6, 11.

3 For more on the traditional understanding of God, see Weinandy, *Does God Suffer?*; Lewis Ayres, "(Mis)Adventures in Trinitarian Ontology," in *The Trinity and an Entangled World: Relationality in Physical Science and Theology*, ed. John C. Polkinghorne (Grand Rapids, MI: Eerdmans, 2010), 130–45.

4 For a fuller critique of process theology, see Nancy Frankenberry, "Some Problems in Process Theodicy," *Religious Studies* 17, no. 2 (1981): 179–197; Foley, *Critique*; Robert C. Neville, *Creativity and God: A Challenge to Process Theology* (New York: Seabury Press, 1980).

EPILOGUE

A DEEPER ENGAGEMENT BETWEEN SCIENCE AND THEOLOGY

Second, as we have seen, evolutionary theology is often in reality far less scientific than it claims to be. Rather than basing their assertions on the strict science of evolution, theologians tend to rely upon broader, philosophical forms of evolutionism. This is understandable. After all, most evolutionary theologians would agree that a genuinely Christian understanding of evolution requires some sense of teleology, which is absent from the scientific theory as such. Nevertheless, when evolution becomes reducible to a metanarrative of development or teleology—when in other words it becomes decidedly *un*scientific—this gravely undermines evolutionary theologians' own goal of integrating science and theology.

Perhaps the greatest pitfall in this direction is when theologians reduce science to a mere repository of useful metaphors. Today, for example, it has become commonplace to use quantum entanglement as a metaphor that proves, as it were, the intimate relationality of all creatures.[5] Of course, quantum entanglement has nothing whatsoever to do with this. Its usefulness in this regard is merely superficial. Bending a scientific term to point to some entirely unrelated theological claim does not constitute a genuine or fruitful interaction between science and religion. Fulton Sheen refers to this methodology as "the lyricism of science."[6] While such an approach appears to give science due consideration, in actuality it abuses science in order to make it into a would-be proof of nonscientific, metaphysical claims. "The laws and theories of matter," writes Sheen, "thus become the laws and theories of spirit."[7]

Such a tendency shows that the task of dialoguing between science and theology is far from over. Our future goal must be to listen ever more attentively to what science has to say and to face it with a steadfast and sober honesty. Most importantly, evolutionary theologians must be clear about how they utilize science and the ways that their conclusions move beyond its authority. They must be careful not to give the false impression that their theological assertions carry the weight of scientific evidence.

Central to this problem is the tendency to view science in terms of authority. When theologians appeal to science as proof of their theological claims,

---

[5] See for example Diarmuid O'Murchu, *Quantum Theology*, rev. ed. (New York: Crossroad, 2004), 32.

[6] Fulton J. Sheen, *God and Intelligence in Modern Philosophy: A Critical Study in the Light of the Philosophy of Saint Thomas* (Garden City, NY: Image Books, 1958).

[7] Sheen, 279 n. 20.

they do so precisely because science carries such weight in contemporary society. Yet the authority of science in regard to nature must not be misconstrued as a general authority in regard to all matters. Natural science can no more prove the Trinity than the doctrine of the Trinity can prove Einstein's theory of general relativity.

Rather than playing a game of authority, evolutionary theologians should look to science as a hermeneutical tool for deepening our understanding of the theological significance of the world. Science can help us to interpret elements of theology, especially inasmuch as these elements pertain to our experience of concrete realities. Hence, for example, a scientific understanding of the close biological relation between humans and other animals can help us to interpret humanity's significance in a way that does not belittle or merely instrumentalize the other creatures of this world. Certainly a wholesome understanding of the interrelation of all creatures is possible without explicit reference to science, but a scientific understanding can complete this recognition by making it all the more vivid and tangible.

Still, the important task of integrating science and theology poses many serious challenges. Without going into excess detail, highlighting a few of these challenges will help to clarify some of the vital questions that must be posed by the evolutionary theologians of the future.

In the first place, science and theology alike are expanding at an increasingly rapid rate. Within theology, we are seeing new and more diverse frameworks of thought, including the proliferation of contextually conscious forms of theology (e.g., feminist, womanist, Black, US Latino/a) and the incorporation of a broad range of contemporary disciplines, methods, and hermeneutics (e.g., the use of sociology, ethnography, economic theory, or literary analysis). While in some cases these theological developments coincide with the fragmentation of Christian denominations or the decline of centralized patterns of belief, in other cases they signify how developments within cohesive Christian communities have at times allowed for a greater range of variation within a still definable concept of orthodoxy. Within science, rapid discoveries in a variety of fields continue to refine and specify our language and understanding of the processes of biological evolution. The astounding pace of scientific research today makes it all the more difficult for the armchair biologist to maintain a precise and unerring conception of the facts; wading through the overwhelming mire of fresh data often requires an expertise and a keen familiarity with terminology that exceeds the competence of the average nonexpert.

Such a rapid expansion of knowledge places a severe limit on the ability of most people to be truly fluent in either discipline as a whole, let alone in both. While the skilled reader will no doubt be willing to overlook minor issues, in some cases even small errors in understanding science or theology can lead to the formation of large, complex schemas that remain fundamentally flawed. It is far too easy for a theologian to fail to recognize the difference between Darwinian natural selection and Mendelian genetics or for a scientist to draw reductionistic conclusions about the supposed real meaning of Christian beliefs without any reference to the actual, lived experience of particular faith communities.

Second, such mutual complexification goes hand in hand with an abundance of technical terminologies and precise distinctions. An astute biologist can find much to criticize about the particular language and emphases employed by evolutionary theologians. For example, it is not uncommon for nonexperts to characterize evolution as driven by simple chance or to characterize natural selection as focused on the formation of new species rather than the propagation of beneficial traits. It would perhaps be too much to demand that an evolutionary theologian be a true expert in both science and religion; and yet, at the very least, fruitful development in this field requires a level of understanding that carefully and sympathetically takes into account the most important and formative concepts of each unique discipline and appreciates them on their own terms. As we have seen, the word "matter," for example, has very different usages within science and within theology, such that statements about matter from the point of view of science are not simply applicable, even after a few modifications, to what theology traditionally refers to as matter.

In view of such differences, an adequate understanding of each field may often require the ability to navigate between fundamentally different conceptual frameworks, without which the claims of each discipline cannot be rightly understood. In the fourth century, Augustine described the pivotal difference it made when he realized that his mental image of spirit (or "mind") was all too material; he had pictured it not as something really transcendent but rather as a kind of thin, diffusive matter like air.[8] The traditional Christian understanding of God and the soul relies upon a radically immaterial concept of spirit, which often did not exist among the various heresies and pagan religions with which Christianity entered into dialogue. In much the same way, the modern,

---

8 Augustine, *Confessions* V, 10, 20.

scientific theory of evolution relies upon certain dramatic conceptual shifts, which can at times be so stark as to give the sense of a religious conversion. Darwinian evolution, for example, does not see individual organisms as mere instances of a universal type or Idea, but as members of a local population of organisms, which is continually shifting and adapting to environmental factors that press upon its long-term survival. It is nevertheless not uncommon in the popular imagination to misconstrue the science of evolution in one direction or the other—either as primarily having to do with the formation of new species or as a matter of the miraculous change of single individuals.

Third, when science is properly allowed to speak on its own terms, the reality is that such terms often provide few concrete avenues for direct theological application. The truth is that quantum entanglement may have nothing significant to offer to theology over and above the more general principles of quantum physics. Most quantum phenomena, after all, are mere embodiments or empirical manifestations of quantum theory in general. Nevertheless, the usefulness of science for theological understanding must not be judged by its ability to prove theological claims or even to provide handy metaphors. Both approaches misunderstand and abuse science and theology alike.

Just as an authentic dialogue between persons requires each to allow the other to speak, so also does an authentic dialogue between science and theology necessitate that each discourse be permitted to decide its own meaning and limitations. Yet this makes the task of dialogue exceptionally difficult. Certainly scientific discoveries can often shed new light on issues of morality, but just as the biological theory of evolution makes no direct claims about God, so also does Christianity's traditional metaphysical understanding of God stand constitutively outside the realms of biology, physics, sociology, and every other discipline of natural and social science. Process theology circumvents the latter limitation by subverting the traditional understanding of God. This move, however, does not of itself legitimize forcing science to serve as proof for metaphysical claims that lie far beyond its own proper sphere.

RESTRUCTURING THE QUESTION ABOUT THEODICY

We turn now to our third key consideration. As we have seen, the problem of theodicy has become the critical starting point for many evolutionary theologians. However, following the pattern of Milton's *Paradise Lost*, such approaches have become obsessed with justifying evil by somehow or other projecting its origin onto God. They argue that God created nature in such a way that evil would be inevitable, that God is incapable or unwilling to alleviate evil, or even that evil ultimately adds up to good. Some such as Jürgen Moltmann bolster such ineffectual claims by appealing to mythological notions of God suffering within God's own self. However academically appealing such responses may be, they provide little to no comfort to those who suffer more than their fair share of the world's evils. Moreover, such abstract notions have no relation to the lived faith of ordinary Christians who turn to the Father, the Son, and the Holy Spirit in the midst of unjust suffering.

Just as the Gospel comes to us in four accounts, the answer to the problem of evil cannot be reduced to one tidy response. This means in particular that we must resist the urge to treat natural evil and moral evil in exactly the same way. Any theodicy that attempts to solve both at the same time simply collapses one into the other, either by making moral evil a matter of natural necessity or else by making natural evil the product of individual freedom. Neither solution is very helpful.

Theodicy cannot be set aside, but neither should it become the absolute starting point of theological inquiry. Christianity is first and foremost a relationship and only secondarily an explanation for the condition of the world. The doctrine of original sin speaks first and foremost about God's loving plan for our salvation and our need for divine grace; only secondarily does it provide any sort of theodicy.

In view of this, any genuine Christian theodicy must ultimately open up in receptivity unto the mystery of God. Process theodicy, in contrast, becomes shallow because of its pretense at depth. By attempting to explain away all evil as the coming-to-be of the good, it provides an easy answer and a premature conclusion to inquiry. To be clear, it is not that there is no answer to the problem of evil. Rather, the case is that the answers we find are too numerous, too complex, and too rich to be distilled into a tidy system.

I find myself drawn to the approach of Johann Baptist Metz, who frames the question of theodicy as an open-ended quest rooted in the real, concrete suffering of real people. Rather than constructing a mythological narrative of divine suffering "in" God, Metz characterizes the struggle of theodicy as *leiden an Gott*, "suffering on account of God."[9] He intentionally uses the same

---

9 Johann Baptist Metz, *Memoria passionis: ein provozierendes Gedächtnis in pluralistischer Gesellschaft* (Freiburg: Herder, 2006), 17–24.

phrase by which one might suffer a disease, e.g., *leiden an Grippe* might be translated as "languishing with the flu." Those who suffer are right to seek an accounting, and only God can provide the ultimate answer. The final and definitive answer to suffering exceeds our understanding in this life, but the quest for theodicy thus becomes in itself part of that answer, a part of how we find meaning *despite* the evils of this world.

As Cynthia Crysdale and Neil Ormerod note, "In the end the problem of evil is a practical problem requiring a practical solution. This practical solution lies at the heart of the Gospel."[10] We are called by grace to become part of the history of God's loving response to the problem of evil. The evangelical imperative of love is theodicy is action; it is how we establish ourselves as part of the solution to evil's essentially meaningless existence.

EVOLVING THE FUTURE OF THEOLOGY

In conclusion, evolutionary theology remains an ongoing task, a mission, and an invitation. There is much more ground to cover and new connections to explore. The goal of this book has been to open up new vistas by exploring the ground already covered. Now it is time for others to take up the task of furthering this field.

Importantly, the future of evolutionary theology need not be confined to a mere reduplication of earlier types. The potential insights of this field are not limited to the patterns set by Teilhard and Whitehead. There is much to be gleaned from these thinkers, but excessive reliance can become a severe limitation.

In particular, I would suggest that evolutionary theology should further pursue three particular areas: Christocentricity, historicity, and grace.

There is more than one way to be Christocentric, and this emphasis is key to many evolutionary theologies. In some sense, from a Christian point of view, the evolutionary progress of the world ties into the role of Jesus Christ both as the divine Wisdom in whom the world is designed and created and as the Jewish man who, by his very profound existence, confirms the meaning and direction of the history of the world. However tempting it may be to universalize theology by abstracting it from Christ, the truth is that for the Christian tradition Christ is the true principle of universality. All humans are called by the love of God precisely because all are one in Jesus Christ.

---

10 Crysdale and Ormerod, *Creator God, Evolving World*, 101.

In fact, the once-and-for-all character of the incarnation points us to the real and inescapable significance of historicity. If we take evolution seriously, it reminds us that what happens in history—even seemingly incidental and minute biological mutations—carries dramatic consequences for the future. Original sin impacts all of us precisely because we are all part of one unified history of the human race. In the same way, our contemporary political, moral, technological, and especially environmental crises all stem from real, historical developments. We too easily write off the small things; fumes, exhaust, and vapor seem too inconsequential to really make a difference. Yet the truth is that we evolve the future of our world through the little decisions that we make, overlook, and excuse. An effective evolutionary theology must call our attention to the consequences of human history while summoning us to build that history anew here and now. The conscientious actions that we undertake today are building the world of tomorrow.

Yet the eschatological tomorrow, which is the genuine object of Christian hope, can never be the mere product of human efforts alone. In the end, what we need most of all is grace. We require God's loving help in order to make a lasting, positive difference that will transcend the mere ebb and flow of time and the back-and-forth struggle of evolutionary self-interests. This means, however, that evolutionary theologians must resist any and every urge to naturalize grace. We must not reduce the Holy Spirit to a mere cog in the natural process of biological evolution. Grace is not something merely given to us as a guaranteed part of our nature. Rather, it is a supernatural gift that transforms, elevates, and completes nature, allowing it to achieve an end that lies beyond its natural capabilities. Most importantly, looked at from another angle, grace *is* that end in itself, for grace is the indwelling of the Holy Spirit and an intimate relationship with the Triune God.

In the end, to the extent that we undertake a genuine *theology*, our exploration will never be over and done. The scientific doctrine of evolution seeks to answer a fascinating conundrum, but evolutionary theology seeks in faith to understand far more: a *mystery*. We thirst to understand in new and ever deeper ways that profound mystery that is the relationship of the Father, the Son, and the Holy Spirit toward this ever-evolving, constantly shifting world. Far from being incidental, this is a relationship that constitutes our very being as creatures and which opens up in awe-filled contemplation unto the mystery of God's own self.

Works Cited

Aberth, John. *An Environmental History of the Middle Ages: The Crucible of Nature*. London: Routledge, 2013.

Acta et decreta concilii provinciae Coloniensisin civitate Coloniensi anno domini MDCCCLX pontificatus Pii PP. IX. decimoquarto celebrati. Cologne: John Peter Bachem, 1862.

Aristotle. *History of Animals*. Translated by D.M. Balme. Vol. 3. Loeb Classical Library 439. Cambridge, MA: Harvard University Press, 1991.

Articulo, Archimedes C. "Towards an Ethics of Technology: Re-Exploring Teilhard de Chardin's Theory of Technology and Evolution." *Open Journal of Philosophy* 4, no. 4 (November 3, 2014): 518–530.

Artigas, Mariano, Thomas F. Glick, and Rafael A. Martínez. *Negotiating Darwin: The Vatican Confronts Evolution, 1877–1902*. Medicine, Science, and Religion in Historical Context. Baltimore: Johns Hopkins University Press, 2006.

Augustine. "The Grace of Christ and Original Sin." In *Answer to the Pelagians*, translated by Roland J. Teske, 384–463. The Works of Saint Augustine: A Translation for the 21st Century I/23. Hyde Park, NY: New City, 1997.

———. *The Literal Meaning of Genesis*. Translated by John Hammond Taylor. Ancient Christian Writers 41–42. New York: Newman, 1982.

Ayala, Francisco J. *Darwin's Gift to Science and Religion*. Washington, DC: Joseph Henry, 2007.

———. "The Myth of Eve: Molecular Biology and Human Origins." *Science* 270, no. 5244 (December 22, 1995): 1930–36.

Ayres, Lewis. "(Mis)Adventures in Trinitarian Ontology." In *The Trinity and an Entangled World: Relationality in Physical Science and Theology*, edited by John C. Polkinghorne, 130–45. Grand Rapids, MI: Eerdmans, 2010.

Bacon, Francis. *De augmentis scientiarum libri IX*. Amsterdam: Henrik Wetstein, 1694.

———. *Sylva Sylvarum: Or, a Natural History in Ten Centuries*. 6th ed. London: William Lee, 1651.

Baltazar, Eulalio R. *Teilhard and the Supernatural*. Baltimore: Helicon, 1966.

Bannister, Robert C. "'The Survival of the Fittest is Our Doctrine': History or Histrionics?" In *Herbert Spencer: Critical Assessments*, edited by John Offer, 2:165–85. New York: Routledge, 2000.

Barbour, Ian G. *Religion in an Age of Science*. 1st ed. San Francisco: Harper and Row, 1990.

———. "Teilhard's Process Metaphysics." *The Journal of Religion* 49, no. 2 (1969): 136–59.

Barker, Ernest. "The Scientific School: Herbert Spencer *and* After Spencer." In *Herbert Spencer: Critical Assessments*, edited by John Offer, 4:5–32. New York: Routledge, 2000.

Baumgartner, Charles. *Le péché originel*. Paris: Desclée de Brouwer, 1969.

Berdyaev, Nikolai. *The Meaning of the Creative Act*. Translated by Donald A. Lowrie. New York: Collier, 1962.

Bergson, Henri. *Creative Evolution*. Translated by Arthur Mitchell. New York: The Modern Library, 1944.

Birch, Charles, and John B. Cobb Jr. *The Liberation of Life: From the Cell to the Community*. Cambridge: Cambridge University Press, 1981.

Blázquez Paniagua, Francisco. "La recepción del darwinismo en la universidad española (1939–1999)." *Anuario de Historia de la Iglesia* 18 (2009): 55–68.

Bonaventure. *The Soul's Journey into God; The Tree of Life; The Life of St. Francis*. Translated by Ewert H. Cousins. Classics of Western Spirituality. New York: Paulist, 1978.

Bonifazi, Conrad. *The Soul of the World: An Account of the Inwardness of Things*. Washington, DC: University Press of America, 1978.

Bowler, Peter J. "Edward Drinker Cope and the Changing Structure of Evolutionary Theory." *Isis* 68, no. 2 (June 1977): 249–65.

———. *Evolution: The History of an Idea*. 25th anniv. ed. Berkeley: University of California Press, 2009.

———. *Progress Unchained: Ideas of Evolution, Human History and the Future*. Cambridge: Cambridge University Press, 2021.

Brachtendorf, Johannes. "The Goodness of Creation and the Reality of Evil: Suffering as a Problem in Augustine's Theodicy." *Augustinian Studies* 31, no. 1 (2000): 79–92.

Braeckman, Antoon. "Whitehead and German Idealism: A Poetic Heritage." *Process Studies* 14, no. 4 (1985): 265–86.

Bresch, Carsten. *Zwischenstufe Leben: Evolution ohne Ziel?* Fischer-Taschenbücher 6802. Frankfurt: Fischer, 1979.

Brown, Robert F. "The Transcendental Fall in Kant and Schelling." *Idealistic Studies* 14 (1984): 49–66.

Bruteau, Beatrice. *God's Ecstasy: The Creation of a Self-Creating World*. New York: Crossroad, 1997.

Bryan, William Jennings. "God and Evolution." *New York Times*, February 26, 1922.

Bulgakov, Sergius. *The Bride of the Lamb*. Translated by Boris Jakim. Grand Rapids, MI: Eerdmans, 2002.

Burchfield, Joe D. "The Age of the Earth and the Invention of Geological Time." *Lyell: The Past is the Key to the Present (Geological Society, London, Special Publications)* 143, no. 1 (1998): 137–43.

Burns, J. Patout. *Theological Anthropology*. Sources of Early Christian Thought. Philadelphia: Fortress Press, 1981.

Calvin, Jean. *Institutes of the Christian Religion*. Edited by John T. McNeill. Library of Christian Classics 20–21. Philadelphia: Westminster, 1960.

Cann, Rebecca L., Mark Stoneking, and Allan C. Wilson. "Mitochondrial DNA and Human Evolution." *Nature* 325, no. 6099 (January 1987): 31–36.

Chen, Jiayan, Weijie Lv, Xueli Zhang, et al. "Animal Age Affects the Gut Microbiota and Immune System in Captive Koalas (Phascolarctos cinereus)." *Microbiology Spectrum* 11, no. 1 (January 5, 2023): e04101-22.

Clark, Linda L. "Social Darwinism in France." *The Journal of Modern History* 53, no. 1 (1981): D1025–D1044.

Clark, Mary T. "The Divine Milieu in Philosophical Perspective." *The Downside Review* 80, no. 258 (1962): 12–25.

Cobb, John B. Jr. *Is It Too Late?: A Theology of Ecology*. Denton, TX: Environmental Ethics Books, 1995.

Cobb, John B. Jr., and David Ray Griffin. *Process Theology: An Introductory Exposition*. Philadelphia: Westminster Press, 1976.

Cole-Turner, Ronald. "Religion, Technology, and the 'Future' of Evolution." In *Evolutionstheorie Und Schöpfungsglaube: Neue Perspektiven Der Debatte*, edited by Hubert Philipp Weber and Rudolf Langthaler, 281–300. Vienna: V and R unipress, 2013.

———. "Transhumanism and Christianity." In *Transhumanism and Transcendence: Christian Hope in an Age of Technological Enhancement*, edited by Ronald Cole-Turner, 193–203. Washington, DC: Georgetown University Press, 2011.

Coon, Carleton S. *The Origin of Races*. 1st ed. New York: Knopf, 1962.

Couenhoven, Jesse. "St. Augustine's Doctrine of Original Sin." *Augustinian Studies* 36, no. 2 (2005): 359–96.

Cousins, Ewert H. *Process Theology: Basic Writings*. New York: Newman, 1971.

Cowdin, Daniel. "The Moral Status of Otherkind in Christian Ethics." In *Christianity and Ecology: Seeking the Well-Being of Earth and Humans*, edited by Dieter T. Hessel and Rosemary Radford Ruether, 261–90. Religions of the World and Ecology. Cambridge, MA: Harvard University Press, 2000.

Crysdale, Cynthia S. W., and Neil Ormerod. *Creator God, Evolving World*. Minneapolis, MN: Fortress Press, 2013.

Daly, Gabriel. *Creation and Redemption*. Theology and Life 25. Wilmington, DE: M. Glazier, 1989.

Darwin, Charles. *Darwin and Women: A Selection of Letters*. Edited by Samantha Evans. Cambridge: Cambridge University Press, 2017.

———. *Darwin: The Indelible Stamp: The Evolution of an Idea*. Edited by James D. Watson. Philadelphia: Running Press, 2005.

———. *More Letters of Charles Darwin: A Record of His Work in a Series of Hitherto Unpublished Letters*. Edited by Francis Darwin and A.C. Seward. 2 vols. London: Murray, 1903.

———. *On the Origin of Species by Means of Natural Selection, or the Preservation of Favoured Races in the Struggle for Life*. 5th ed. London: Murray, 1869.

———. *The Autobiography of Charles Darwin, 1809–1882*. Edited by Nora Barlow. New York: Norton, 1969.

———. *The Life and Letters of Charles Darwin*. Edited by Francis Darwin. 3 vols. 2nd ed. London: Murray, 1887.

———. *The Origin of Species by Means of Natural Selection, or the Preservation of Favoured Races in the Struggle for Life*. 6th ed. London: Murray, 1872.

Dawkins, Richard. *The Selfish Gene*. 30th anniv. ed. Oxford: Oxford University Press, 2006.

De Laguna, Grace A. "Being and Knowing: A Dialectical Study." *The Philosophical Review* 45, no. 5 (1936): 435–456.

De Vany, Arthur. *The New Evolution Diet: What Our Paleolithic Ancestors Can Teach Us About Weight Loss, Fitness, and Aging.* Emmaus, PA: Rodale, 2011.

Deane-Drummond, Celia. *A Primer in Ecotheology: Theology for a Fragile Earth.* Eugene, OR: Cascade Books, 2017.

———. "Believing Deeply in Creation: Christ and Evolution as Theodrama." In *Evolutionstheorie Und Schöpfungsglaube: Neue Perspektiven Der Debatte*, edited by Hubert Philipp Weber and Rudolf Langthaler, 187–200. Vienna: V and R unipress, 2013.

———. *Christ and Evolution: Wonder and Wisdom.* Theology and the Sciences. Minneapolis, MN: Fortress Press, 2009.

———. "Future Perfect? God, the Transhuman Future and the Quest for Immortality." In *Future Perfect?: God, Medicine and Human Identity*, edited by Celia Deane-Drummond and Peter Scott, 168–82. London: T and T Clark International, 2006.

———. "In Adam All Die?: Questions at the Boundary of Niche Construction, Community Evolution, and Original Sin." In *Evolution and the Fall*, edited by William T. Cavanaugh and James K. A. Smith, 23–47. Grand Rapids, MI: Eerdmans, 2017.

———. *The Wisdom of the Liminal: Evolution and Other Animals in Human Becoming.* Grand Rapids, MI: Eerdmans, 2014.

Delio, Ilia. *The Unbearable Wholeness of Being: God, Evolution and the Power of Love.* Maryknoll, NY: Orbis, 2013.

———. "Transhumanism or Ultrahumanism?: Teilhard de Chardin on Technology, Religion and Evolution." *Theology and Science* 10, no. 2 (2012): 153–166.

Depoortere, Frederiek. "'God Himself Is Dead!' Luther, Hegel, and the Death of God." *Philosophy and Theology* 19, no. 1/2 (July 1, 2007): 171–195.

Dobzhansky, Theodosius, Ashley Montagu, and C.S. Coon. "Two Views of Coon's *Origin of Races* with Comments by Coon and Replies." *Current Anthropology* 4, no. 4 (1963): 360–67.

Domning, Daryl P. *Original Selfishness: Original Sin and Evil in the Light of Evolution.* Ashgate Science and Religion. Aldershot, England: Ashgate, 2006.

Dorrien, Gary J. *In a Post-Hegelian Spirit: Philosophical Theology as Idealistic Discontent.* Waco, TX: Baylor University Press, 2020.

Drozdov, Filaret. *Select Sermons.* London: J. Masters, 1873.

———. "The Longer Catechism of the Russian Church." In *The Doctrine of the Russian Church: Being the Primer or Spelling Book, the Shorter and Longer Catechisms, and a Treatise on the Duty of Parish Priests*, 28–142. London: J. Masters, 1845.

Duffy, Stephen. "Genes, Original Sin and the Human Proclivity to Evil." *Horizons* 32, no. 2 (2005): 210–34.

Dupree, A. Hunter. "Christianity and the Scientific Community in the Age of Darwin." In *God and Nature: Historical Essays on the Encounter Between Christianity and Science*, edited by David C. Lindberg and Ronald L. Numbers, 351–68. Berkeley: University of California Press, 1986.

Eckhart, Meister. *Meister Eckhart: Teacher and Preacher.* Translated by Bernard McGinn. Classics of Western Spirituality. New York: Paulist Press, 1986.

Edwards, Denis. *The God of Evolution: A Trinitarian Theology.* New York: Paulist, 1999.

Eliade, Mircea. *The Myth of the Eternal Return: Cosmos and History.* 2nd pbk. ed. Bollingen 46. Princeton, NJ: Princeton University Press, 2005.

Ellis, Robert. "From Hegel to Whitehead." *The Journal of Religion* 61, no. 4 (1981): 403–421.

Empedocles. *The Poem of Empedocles: A Text and Translation with an Introduction.* Translated by Brad Inwood. Rev. ed. Toronto: University of Toronto Press, 2001.

Feuerbach, Ludwig. *Principles of the Philosophy of the Future.* Translated by Manfred H. Vogel. The Library of Liberal Arts 197. Indianapolis: Bobbs-Merrill, 1966.

Fiske, John. *Through Nature to God.* Boston: Houghton Mifflin, 1899.

Fitzpatrick, Joseph. *The Fall and the Ascent of Man: How Genesis Supports Darwin.* Lanham, MD: University Press of America, 2012.

Foley, Leo Albert. *A Critique of the Philosophy of Being of Alfred North Whitehead in the Light of Thomistic Philosophy.* Philosophical studies (Catholic University of America) v. 94. Washington, DC: Catholic University of America Press, 1946.

Frankenberry, Nancy. "Some Problems in Process Theodicy." *Religious Studies* 17, no. 2 (1981): 179–197.

Galton, Francis. *Inquiries into Human Faculty and Its Development.* 2nd ed. London: J.M. Dent, 1911.

Gilbert, W.S. *Princess Ida: Or, Castle Adamant.* London: G. Bell, 1912.

Gilson, Étienne. *God and Philosophy.* New Haven, CT: Yale University Press, 1941.

Glass, Bentley. "The Germination of the Idea of Biological Species." In *Forerunners of Darwin, 1745–1859,* edited by Bentley Glass, Owsei Temkin, and William L. Straus, 30–48. Baltimore: Johns Hopkins University Press, 1968.

Gould, Stephen Jay. *Ever Since Darwin: Reflections in Natural History.* New York: Norton, 1977.

Gregersen, Niels Henrik. "The Cross of Christ in an Evolutionary World." *Dialog* 40, no. 3 (2001): 192–207.

Griffin, David Ray, John B. Cobb Jr., and Clark H. Pinnock, eds. *Searching for an Adequate God: A Dialogue Between Process and Free Will Theists.* Grand Rapids, MI: Eerdmans, 2000.

Grumett, David. *Christ in the World of Matter: Teilhard de Chardin's Religious Experience and Vision.* Teilhard Studies 67. Woodbridge, CT: American Teilhard Association, 2013.

———. "Metaphysics, Morality, and Politics." In *From Teilhard to Omega: Co-Creating an Unfinished Universe.* Edited by Ilia Delio, 111–26. Maryknoll, NY: Orbis, 2014.

———. *Teilhard de Chardin: Theology, Humanity and Cosmos.* Studies in Philosophical Theology 29. Leuven: Peeters, 2005.

Guralnick, Stanley M. "Geology and Religion before Darwin: The Case of Edward Hitchcock, Theologian and Geologist (179–864)." *Isis* 63, no. 4 (1972): 529–43.

Gutiérrez, Gustavo. *Las Casas: In Search of the Poor of Jesus Christ.* Maryknoll, NY: Orbis, 1993.

Haight, Roger. *Faith and Evolution: A Grace-Filled Naturalism.* Maryknoll, NY: Orbis, 2019.

Hale, Robert. *Christ and the Universe: Teilhard de Chardin and the Cosmos.* Chicago: Franciscan Herald, 1973.

Harnack, Adolf von. *History of Dogma.* 7 vols. London: Williams and Norgate, 1894.

Hartshorne, Charles. "Introduction: The Development of Process Philosophy." In *Philosophers of Process,* edited by Douglas Browning, v–xxii. New York: Random House, 1965.

Haught, John F. *Making Sense of Evolution: Darwin, God, and the Drama of Life.* Louisville, KY: Westminster John Knox Press, 2010.

Hawks, John, and Milford H. Wolpoff. "Sixty Years of Modern Human Origins in the American Anthropological Association." *American Anthropologist* 105, no. 1 (2003): 89–100.

Hefner, Philip J. "Sociobiology, Ethics, and Theology." *Zygon* 19, no. 2 (1984): 185–207.

———. *Technology and Human Becoming*. Facets. Minneapolis, MN: Fortress Press, 2003.

———. *The Human Factor: Evolution, Culture, and Religion*. Theology and the Sciences. Minneapolis, MN: Fortress Press, 1993.

Hegel, Georg Wilhelm Friedrich. *G. W. F. Hegel: Theologian of the Spirit*. Edited by Peter C. Hodgson. Minneapolis, MN: Fortress Press, 1997.

———. *Hegel's Philosophy of Mind*. Translated by William Wallace. Oxford: Clarendon, 1894.

———. *Hegel's Science of Logic*. Translated by Arnold V. Miller. Amherst, NY: Humanity Books, 1999.

———. *Lectures on the Philosophy of History*. Translated by Ruben Alvarado. Aalten, Netherlands: WordBridge, 2011.

———. *Lectures on the Philosophy of World History: Introduction, Reason in History*. Translated by H.B. Nisbet. Cambridge Studies in the History and Theory of Politics. Cambridge: Cambridge University Press, 1975.

———. *Phenomenology of Spirit*. Translated by Arnold V. Miller. Oxford: Clarendon, 1977.

———. *The Logic of Hegel*. Translated by William Wallace. 2nd ed. Oxford: Clarendon Press, 1892.

Herbert, Sandra. "Darwin, Malthus, and Selection." *Journal of the History of Biology* 4, no. 1 (1971): 209–17.

Hessel, Dieter T., and Rosemary Radford Ruether, eds. *Christianity and Ecology: Seeking the Well-being of Earth and Humans*. Religions of the World and Ecology. Cambridge, MA: Harvard University Press, 2000.

Hick, John. *Death and Eternal Life*. New York: Harper and Row, 1976.

———. *Evil and the God of Love*. 1st ed. New York: Harper and Row, 1966.

Horn, Stephan Otto, and Siegfried Wiedenhofer, eds. *Creation and Evolution: A Conference with Pope Benedict XVI in Castel Gandolfo*. Translated by Michael J. Miller. San Francisco: Ignatius, 2008.

Hulsbosch, Ansfridus. *God in Creation and Evolution*. Translated by Martin Versfeld. New York: Sheed and Ward, 1966.

Hunter, Cornelius G. *Darwin's God: Evolution and the Problem of Evil*. Grand Rapids, MI: Brazos, 2001.
Huxley, Julian. *Evolution: The Modern Synthesis*. New York: Harper and Brothers, 1942.
———. "Transhumanism." In *New Bottles For New Wine*, 13–17. London: Chatto and Windus, 1957.
Huxley, T.H. "The Genealogy of Animals." In *Critiques and Addresses*, 303–19. London: Macmillan, 1873.
Illingworth, John Richardson. "The Incarnation in Relation to Development." In *Lux Mundi: A Series of Studies in the Religion of the Incarnation*, edited by Charles Gore, 132–57. 15th ed. London: J. Murray, 1904.
Jenkins, Willis. "After Lynn White: Religious Ethics and Environmental Problems." *The Journal of Religious Ethics* 37, no. 2 (2009): 283–309.
Johnson, Rebecca N., Denis O'Meally, Zhiliang Chen, et al. "Adaptation and Conservation Insights from the Koala Genome." *Nature Genetics* 50, no. 8 (August 2018): 1102–11.
Jonas, Hans. *The Gnostic Religion: The Message of the Alien God and the Beginnings of Christianity*. 3rd ed. Boston: Beacon, 2001.
Kant, Immanuel. *Critique of Pure Reason*. Translated by J.M.D. Meiklejohn. London: George Bell and Sons, 1887.
———. "Lectures on the Philosophical Doctrine of Religion." In *Religion and Rational Theology*, edited by George Di Giovanni, translated by Allen W. Wood, 335–452. The Cambridge Edition of the Works of Immanuel Kant. Cambridge: Cambridge University Press, 1996.
———. "On the Miscarriage of All Philosophical Trials in Theodicy." In *Religion and Rational Theology*, edited by Allen W. Wood, translated by George Di Giovanni, 19–38. The Cambridge Edition of the Works of Immanuel Kant. Cambridge: Cambridge University Press, 1996.
———. "Religion Within the Boundaries of Mere Reason." In *Religion and Rational Theology*, edited by Allen W. Wood, translated by George Di Giovanni, 39–216. The Cambridge Edition of the Works of Immanuel Kant. Cambridge: Cambridge University Press, 1996.
Kaplan, Jed O., Kristen M. Krumhardt, Erle C. Ellis, William F. Ruddiman, Carsten Lemmen, and Kees Klein Goldewijk. "Holocene Carbon Emissions as a Result of Anthropogenic Land Cover Change." *The Holocene* 21, no. 5 (August 1, 2011): 775–791.

Kasujja, Augustine. *Polygenism and the Theology of Original Sin: Eastern African Contribution to the Solution of the Scientific Problem, The Impact of Polygenism in Modern Theology*. Collectio Urbaniana 3268. Rome: Urbaniana University Press, 1986.

Keller, Evelyn Fox. *The Century of the Gene*. Cambridge, MA: Harvard University Press, 2002.

Kemp, Kenneth W. "Science, Theology, and Monogenesis." *American Catholic Philosophical Quarterly* 85, no. 2 (2011): 217–36.

Kenny, Robert. "From the Curse of Ham to the Curse of Nature: The Influence of Natural Selection on the Debate on Human Unity before the Publication of *The Descent of Man*." *The British Journal for the History of Science* 40, no. 3 (2007): 367–88.

Kepes, Jr., Theodore. "Toward a Unified Vision: The Integration of Christian Theology and Evolution in Karl Rahner's Understanding of Matter and Spirit." *Philosophy and Theology* 20, no. 1–2 (2008): 269–90.

Klauder, Francis J. *Aspects of the Thought of Teilhard de Chardin*. North Quincy, MA: Christopher Pub. House, 1971.

Kline, George L. "Concept and Concrescence: An Essay in Hegelian-Whiteheadian Ontology." In *Hegel and Whitehead: Contemporary Perspectives on Systematic Philosophy*, edited by George R. Lucas Jr., 133–51. Albany, NY: State University of New York Press, 1986.

Korsmeyer, Jerry D. *Evolution and Eden: Balancing Original Sin and Contemporary Science*. New York: Paulist, 1998.

LaCugna, Catherine Mowry. *God for Us: The Trinity and Christian Life*. San Francisco: HarperSanFrancisco, 1991.

Lamarck, Jean Baptiste. *Zoological Philosophy: An Exposition with Regard to the Natural History of Animals*. New York: Hafner, 1963.

Leibniz, Gottfried Wilhelm. *Leibniz's Monadology: A New Translation and Guide*. Translated by Lloyd Strickland. Edinburgh: Edinburgh University Press, 2014.

———. *New Essays Concerning Human Understanding*. Translated by Alfred G. Langley. New York: Macmillan, 1896.

———. *Theodicy: Essays on the Goodness of God, the Freedom of Man, and the Origin of Evil*. Edited by Austin Farrer. London: Routledge, 1952.

Linné, Carl von. *Philosophia botanica*. Stockholm: G. Kiesewetter, 1751.

Liu, Yongsheng. "Darwinian Evolution Includes Lamarckian Inheritance of Acquired Characters." *International Journal of Epidemiology* 45, no. 6 (December 1, 2016): 2206–2207.

Livingstone, David N. *Darwin's Forgotten Defenders: The Encounter between Evangelical Theology and Evolutionary Thought*. Grand Rapids, MI: Eerdmans, 1987.

Lovejoy, Arthur O. *The Great Chain of Being: A Study of the History of an Idea*. Cambridge, MA: Harvard University Press, 1936.

Lowe, Victor. *Understanding Whitehead*. Baltimore: Johns Hopkins Press, 1962.

de Lubac, Henri. *Teilhard de Chardin: The Man and His Meaning*. 1st American ed. New York: Hawthorn Books, 1965.

———. *The Religion of Teilhard de Chardin*. New York: Desclee, 1967.

Lucas Jr., George R. "Evolutionist Theories and Whitehead's Philosophy." *Process Studies* 14 (December 1, 1985): 287–300.

———. *Two Views of Freedom in Process Thought: A Study of Hegel and Whitehead*. Dissertation Series—American Academy of Religion 28. Missoula, MT: Scholars, 1979.

Lukas, Mary, and Ellen Lukas. *Teilhard*. Garden City, NY: Doubleday, 1977.

Luther, Martin. *Early Theological Works*. Translated by James Atkinson. Library of Christian Classics 16. Philadelphia: Westminster, 1962.

Lyon, John. "Immediate Reactions to Darwin: The English Catholic Press' First Reviews of the 'Origin of the Species.'" *Church History* 41, no. 1 (1972): 78–93.

Mahoney, John. *Christianity in Evolution: An Exploration*. Washington, DC: Georgetown University Press, 2011.

Malthus, Thomas Robert. *First Essay on Population, 1798*. Edited by James Bonar. London: Macmillan, 1926.

Maritain, Jacques. *Moral Philosophy: An Historical and Critical Survey of the Great Systems*. London: G. Bles, 1964.

Mastrantonis, George. *A New-Style Catechism on the Eastern Orthodox Faith for Adults*. 2nd ed. St. Louis: The Ologos Mission, 1977.

Mathieu, Pierre-Louis. *La pensée politique at economique de Teilhard de Chardin*. Paris: Éditions du Seuil, 1969.

McFague, Sallie. *A New Climate for Christology: Kenosis, Climate Change, and Befriending Nature*. Minneapolis, MN: Fortress Press, 2021.

———. *Super, Natural Christians: How We Should Love Nature*. Minneapolis, MN: Fortress Press, 1997.

———. *The Body of God: An Ecological Theology*. Minneapolis, MN: Fortress Press, 1993.

McGrath, Alister E. *Darwinism and the Divine: Evolutionary Thought and Natural Theology*. Malden, MA: Wiley-Blackwell, 2011.

McNally, David. "Political Economy to the Fore: Burke, Malthus and the Whig Response to Popular Radicalism in the Age of the French Revolution." *History of Political Thought* 21, no. 3 (2000): 427–47.

Meiners, Christoph. *Grundriss der Geschichte der Menschheit*. 1st ed. Lemgo, Germany: Meyerschen Buchhandlung, 1785.

Mellert, Robert B. *What is Process Theology?* New York: Paulist, 1975.

Mesle, C. Robert. "Does God Hide from Us?: John Hick and Process Theology on Faith, Freedom and Theodicy." *International Journal for Philosophy of Religion* 24, no. 1/2 (1988): 93–111.

———. *Process-Relational Philosophy: An Introduction to Alfred North Whitehead*. Electronic resource. West Conshohocken, PA: Templeton Foundation, 2008.

Metz, Johann Baptist. *Memoria passionis: ein provozierendes Gedächtnis in pluralistischer Gesellschaft*. Freiburg: Herder, 2006.

Midgley, Mary. *Evolution as a Religion: Strange Hopes and Stranger Fears*. Rev. ed. Routledge Classics. London: Routledge, 2002.

Milton, John. *Johann Miltons Verlust des Paradieses; ein Helden-Gedicht*. Translated by Johann Jakob Bodmer. 1st ed. Zurich: M. Rordorf, 1732.

Mivart, St. George Jackson. *Essays and Criticisms*. 2 vols. London: J.R. Osgood, 1892.

———. "Modern Catholics and Scientific Freedom." *The Nineteenth Century* 18 (1885): 30–47.

———. *On the Genesis of Species*. 2nd ed. London: Macmillan, 1871.

Moltmann, Jürgen. *God in Creation: A New Theology of Creation and the Spirit of God*. Translated by Margaret Kohl. Gifford Lectures 1984–85. San Francisco: Harper and Row, 1985.

———. "The 'Crucified God' and the Trinity Today." In *New Questions on God*, translated by David Smith, 26–37. Concilium 76. New York: Herder and Herder, 1972.

———. *The Way of Jesus Christ: Christology in Messianic Dimensions*. 1st HarperCollins ed. San Francisco: HarperSanFrancisco, 1990.

Moore, James R. *The Post-Darwinian Controversies: A Study of the Protestant Struggle to Come to Terms with Darwin in Great Britain and America, 1870–1900*. Cambridge: Cambridge University Press, 1979.

Morrow, Jeffrey L. "French Apocalyptic Messianism: Isaac La Peyrère and Political Biblical Criticism in the Seventeenth Century." *Toronto Journal of Theology* 27, no. 2 (2011): 203–213.

Murphy, George L. "Cosmology, Evolution, and Biotechnology." In *Bridging Science and Religion*, edited by Ted Peters and Gaymon Bennett, 196–212. Theology and the Sciences. Minneapolis, MN: Fortress Press, 2003.

Nemesius. "A Treatise on the Nature of Man." In *Cyril of Jerusalem and Nemesius of Emesa*, edited by William Telfer, 224–453. Library of Christian Classics 4. Philadelphia: Westminster, 1955.

Neville, Robert C. *Creativity and God: A Challenge to Process Theology*. New York: Seabury Press, 1980.

Ngien, Dennis. "Chalcedonian Christology and Beyond: Luther's Understanding of the *Communicatio Idiomatum*." *The Heythrop Journal* 45, no. 1 (2004): 54–68.

Nietzsche, Friedrich Wilhelm. *The Anti-Christ, Ecce Homo, Twilight of the Idols, and Other Writings*. Edited by Aaron Ridley and Judith Norman. Translated by Judith Norman. Cambridge Texts in the History of Philosophy. Cambridge: Cambridge University Press, 2005.

———. *The Gay Science: With a Prelude in German Rhymes and an Appendix of Songs*. Edited by Bernard Williams. Translated by Josefine Nauckhoff and Adrian Del Caro. Cambridge Texts in the History of Philosophy. Cambridge: Cambridge University Press, 2001.

Nisbet, Ebenezer. *The Science of the Day and Genesis*. C. Venton Patterson, 1886.

Nogar, Raymond J. *The Wisdom of Evolution*. Garden City, NJ: Doubleday, 1963.

Northcott, Michael S. *The Environment and Christian Ethics*. New Studies in Christian Ethics. Cambridge: Cambridge University Press, 1996.

Nott, J.C. "The Mulatto a Hybrid—Probable Extermination of the Two Races If the Whites and Blacks Are Allowed to Intermarry." *The Boston Medical and Surgical Journal* 29, no. 2 (August 16, 1843): 29–32.

Olkovich, Nicholas. "Reinterpreting Original Sin: Integrating Insights from Sociology and the Evolutionary Sciences." *Heythrop Journal* 54, no. 5 (2013): 715–31.

O'Murchu, Diarmuid. *Quantum Theology*. Rev. ed. New York: Crossroad, 2004.

Oosthuizen, Jacobus Stefanus. *Van Plotinus tot Teilhard de Chardin. 'n Studie oor die metamorfose van die Westerse werklikheidsbeeld*. Amsterdam: Rodopi, 1974.

O'Regan, Cyril. *The Anatomy of Misremembering: Von Balthasar's Response to Philosophical Modernity*. Chestnut Ridge, NY: Crossroad Pub., 2014.

———. *The Heterodox Hegel*. SUNY Series in Hegelian Studies. Albany: SUNY Press, 1994.

O'Rourke, Fran. "Aristotle and the Metaphysics of Evolution." *Review of Metaphysics* 58, no. 1 (2004): 3–59.

Orr, James. *God's Image in Man and Its Defacement in the Light of Modern Denials*. London: Hodder and Stoughton, 1905.

Peacocke, Arthur. *Evolution: The Disguised Friend of Faith?: Selected Essays*. Philadelphia: Templeton Foundation, 2004.

Pernoud, Régine. *Those Terrible Middle Ages: Debunking the Myths*. Translated by Anne Englund Nash. San Francisco, CA: Ignatius, 2000.

Peters, Ted. "The Evolution of Evil." In *The Evolution of Evil*, edited by Gaymon Bennett, Martinez J. Hewlett, Ted Peters, and Robert John Rossell, 19–52. Religion, Theologie und Naturwissenschaft 8. Göttingen: Vandenhoeck and Ruprecht, 2008.

Peters, Ted, and Martinez J. Hewlett. *Theological and Scientific Commentary on Darwin's Origin of Species*. Nashville, TN: Abingdon, 2008.

Peterson, Gregory R. "Falling Up: Evolution and Original Sin." In *Evolution and Ethics: Human Morality in Biological and Religious Perspective*, edited by Philip Clayton and Jeffrey Schloss, 273–86. Grand Rapids, MI: Eerdmans, 2004.

Pittenger, W. Norman. *Alfred North Whitehead*. Makers of Contemporary Theology. London: Lutterworth, 1969.

Pius XII. *Humani generis*. New York: Paulist, 1950.

Plato. *Laws*. Translated by R.G. Bury. 2 vols. Loeb Classical Library 187, 192. Cambridge, MA: Harvard University Press, 1926.

———. *Phaedrus*. Translated by Robin Waterfield. Oxford World's Classics. Oxford: Oxford University Press, 2002.

Plotinus. *Enneads*. Translated by A. H. Armstrong. 6 vols. Loeb Classical Library 440–45. Cambridge, MA: Harvard University Press, 1966.

Polkinghorne, J.C. *Scientists as Theologians: A Comparison of the Writings of Ian Barbour, Arthur Peacocke and John Polkinghorne*. London: SPCK, 1996.
Quash, Ben. *Theology and the Drama of History*. Cambridge Studies in Christian Doctrine 13. Cambridge: Cambridge University Press, 2005.
Rahner, Karl. "Concerning the Relationship Between Nature and Grace." In *Theological Investigations*, translated by Cornelius Ernst, 1:297–317. Baltimore: Helicon, 1961.
———. "Evolution and Original Sin." In *The Evolving World and Theology*, edited by Johann Baptist Metz, translated by Theodore L. Westow, 61–73. Concilium 26. New York: Paulist, 1967.
———. "Exkurs: Erbsünde und Monogenismus." In *Theologie der Erbsünde*, by Karl-Heinz Weger, 176–223. Quaestiones Disputatae 44. Freiburg: Herder, 1970.
———. *Foundations of Christian Faith: An Introduction to the Idea of Christianity*. Translated by William V. Dych. New York: Crossroad, 1982.
———. *Hominisation: The Evolutionary Origin of Man as a Theological Problem*. Translated by W.J. O'Hara. Quaestiones disputatae 13. New York: Herder and Herder, 1965.
———. *Meditations on Priestly Life*. Translated by Edward Quinn. London: Sheed and Ward, 1973.
———. "The Problem of Genetic Manipulation." In *Theological Investigations*, translated by Graham Harrison, 9:225–52. New York: Herder and Herder, 1972.
———. "The Sin of Adam." In *Theological Investigations*, translated by David Bourke, 11:247–62. New York: Seabury, 1974.
———. "Theological Reflexions on Monogenism." In *Theological Investigations*, translated by Cornelius Ernst, 1:229–96. Baltimore: Helicon, 1961.
———. "Unity of the Church—Unity of Mankind." In *Theological Investigations*, edited by Paul Imhof, translated by Edward Quinn, 20:154–72. New York: Crossroad, 1981.
Rahner, Karl, and Herbert Vorgrimler. *Dictionary of Theology*. Translated by Richard Strachan. 2nd ed. New York: Crossroad, 1985.
Ratzinger, Joseph. "The God of Faith and the God of the Philosophers." In *Introduction to Christianity*, 137–50. Rev. ed. San Francisco: Ignatius, 2004.
Ray, John. *Methodus plantarum nova*. London: Henry Faithorne and John Kersey, 1682.

WORKS CITED

Reeves, Michael, and Hans Madueme. "Threads in a Seamless Garment: Original Sin in Systematic Theology." In *Adam, the Fall, and Original Sin: Theological, Biblical, and Scientific Perspectives*, edited by Hans Madueme and Michael Reeves, 209–24. Grand Rapids, MI: Baker Academic, 2014.

Rigby, Paul. *Original Sin in Augustine's Confessions*. Ottawa: University of Ottawa Press, 1987.

Rogers, James Allen. "Darwinism and Social Darwinism." In *Herbert Spencer: Critical Assessments*, edited by John Offer, 2:149–64. New York: Routledge, 2000.

Rondet, Henri. *Original Sin: The Patristic and Theological Background*. Translated by Cajetan Finegan. Staten Island, NY: Alba House, 1972.

Royer, Clémence. "Préface de la Première Édition." In *De l'origine des espèces par sélection naturelle, ou Des lois de transformation des êtres organisés*, by Charles Darwin, xv–lxxii. 3rd ed. Paris: V. Masson et fils, 1870.

Ruether, Rosemary Radford. *To Change the World: Christology and Cultural Criticism*. New York: Crossroad, 1981.

Santmire, H. Paul. *Nature Reborn: The Ecological and Cosmic Promise of Christian Theology*. Theology and the Sciences. Minneapolis, MN: Fortress Press, 2000.

Savage, Minot J. *Evolution and Religion, from the Standpoint of One Who Believes in Both*. Philadelphia: G.H. Buchanan, 1886.

———. *Religion for To-day*. Boston: G.E. Ellis, 1897.

———. *The Religion of Evolution*. Boston: Lockwood, Brooks and Co., 1876.

Schelling, Friedrich Wilhelm Joseph von. *Of Human Freedom*. Translated by James Gutmann. Chicago: Open Court, 1936.

Schlitzer, Albert. "The Position of Modern Theology on the Evolution of Man." *Laval théologique et philosophique* 8, no. 2 (1952): 208–229.

Schmitz-Moormann, Karl. "The Future of Teilhardian Theology." *Zygon* 30, no. 1 (March 1, 1995): 117–29.

Schönborn, Christoph von. "Foreword." In *Creation and Evolution: A Conference with Pope Benedict XVI in Castel Gandolfo*, edited by Stephan Otto Horn and Siegfried Wiedenhofer, translated by Michael J. Miller and Michael J. Miller, 7–23. San Francisco: Ignatius, 2008.

———. *God's Human Face: The Christ-Icon*. Translated by Lothar Krauth. San Francisco: Ignatius, 1994.

———. *Man, the Image of God: The Creation of Man as Good News*. Translated by Henry Taylor and Michael J. Miller. San Francisco: Ignatius Press, 2011.

Schwager, Raymund. *Banished from Eden: Original Sin and Evolutionary Theory in the Drama of Salvation*. Translated by James Williams. Inigo Texts 9. Leominster, UK: Gracewing, 2006.

Scott, Mark S.M. "Suffering and Soul-Making: Rethinking John Hick's Theodicy." *Journal of Religion* 90, no. 3 (2010): 313–34.

Sheen, Fulton J. *God and Intelligence in Modern Philosophy: A Critical Study in the Light of the Philosophy of Saint Thomas*. Garden City, NY: Image Books, 1958.

Sideris, Lisa H. *Environmental Ethics, Ecological Theology, and Natural Selection*. Columbia Series in Science and Religion. New York: Columbia University Press, 2003.

Simpson, Richard. "Darwin on the Origin of Species." *The Rambler*, March 1860.

Solovyov, Vladimir. "The Idea of a Superman." In *Politics, Law, and Morality*, translated by Vladimir Wozniuk, 255–63. Russian Literature and Thought. New Haven, CT: Yale University Press, 2000.

Spencer, Herbert. *First Principles*. New York: De Witt Revolving Fund, 1958.

———. *The Man Versus the State: With Six Essays on Government, Society, and Freedom*. Indianapolis, IN: Liberty Classics, 1981.

———. *The Principles of Biology*. 2 vols. London: Williams and Norgate, 1864.

Teilhard de Chardin, Pierre. *Activation of Energy*. Translated by René Hague. New York: Harcourt Brace Jovanovich, 1971.

———. *Building the Earth*. Translated by Noël Lindsay. New York: Avon Books, 1969.

———. *Christianity and Evolution*. Translated by René Hague. New York: Harcourt Brace Jovanovich, 1971.

———. *Human Energy*. Translated by J.M. Cohen. New York: Harcourt Brace Jovanovich, 1971.

———. *Science and Christ*. Translated by René Hague. New York: Harper and Row, 1968.

———. *The Future of Man*. New York: Harper and Row, 1964.

———. *The Future of Man*. Translated by Norman Denny. New York: Doubleday, 2004.

———. *The Heart of Matter*. Translated by René Hague. New York: Harcourt Brace Jovanovich, 1979.

———. *The Making of a Mind: Letters from a Soldier-Priest, 1914–1919*. Translated by René Hague. 1st ed. New York: Harper and Row, 1965.

———. *The Phenomenon of Man*. Translated by Bernard Wall. New York: Harper and Brothers, 1959.

———. *Toward the Future*. Translated by René Hague. San Diego: Harcourt, 2002.

Tertullian. *Adversus Marcionem*. Edited by Ernest Evans. Oxford Early Christian texts. Oxford: Clarendon, 1972.

Tyneh, Carl S., ed. *Orthodox Christianity: Overview and Bibliography*. New York: Nova Science Publishers, 2003.

Vallade, Jean. "The Slow Emergence of the Meiosis Concept from 1882 to 1909." *Acta Botanica Gallica* 160, no. 1 (2013): 3–10.

Van den Brink, Gijsbert. "Questions, Challenges, and Concerns for Original Sin." In *Finding Ourselves After Darwin: Conversations on the Image of God, Original Sin, and the Problem of Evil*, edited by Stanley P. Rosenberg, Michael S. Burdett, Michael Lloyd, and Benno van den Toren, 117–29. Grand Rapids, MI: Baker Academic, 2018.

Van den Toren, Benno. "Original Sin and the Coevolution of Nature and Culture." In *Finding Ourselves After Darwin: Conversations on the Image of God, Original Sin, and the Problem of Evil*, edited by Stanley P. Rosenberg, Michael S. Burdett, Michael Lloyd, and Benno van den Toren, 173–86. Grand Rapids, MI: Baker Academic, 2018.

Viney, Donald Wayne. "Teilhard: *Le Philosophe malgré l'Église*." In *Rediscovering Teilhard's Fire*, edited by Kathleen Duffy, 69–88. Philadelphia: Saint Joseph's University Press, 2010.

Vlastos, Gregory. "Organic Categories in Whitehead." *The Journal of Philosophy* 34, no. 10 (1937): 253–262.

Vogt, Karl Christoph. *Lectures on Man: His Place in Creation, and in the History of the Earth*. Edited by James Hunt. London: Longman, et al., 1864.

Wandinger, Nikolaus. *Die Sündenlehre als Schlüssel zum Menschen: Impulse K. Rahners und R. Schwagers zu einer Heuristik theologischer Anthropologie*. Beiträge zur mimetischen Theorie 16. Münster: Lit, 2003.

Weikart, Richard. "The Origins of Social Darwinism in Germany, 1859–1895." *Journal of the History of Ideas* 54, no. 3 (1993): 469–88.

Weinandy, Thomas G. *Does God Suffer?* Notre Dame, IN: University of Notre Dame, 2000.

White, Lynn. "The Historical Roots of Our Ecologic Crisis." *Science* 155, no. 3767 (1967): 1203–1207.

White, Michael J. "Stoic Natural Philosophy (Physics and Cosmology)." In *The Cambridge Companion to the Stoics*, edited by Brad Inwood, 124–52. Cambridge Companions to Philosophy. Cambridge: Cambridge University Press, 2003.

Whitehead, Alfred North. *Adventures of Ideas*. New York: New American Library, 1955.

———. *Nature and Life*. Cambridge Miscellany 13. Cambridge: Cambridge University Press, 1934.

———. *Process and Reality: An Essay in Cosmology*. Edited by David Ray Griffin and Donald W. Sherburne. Corrected ed. New York: The Free Press, 1978.

———. *Religion in the Making: Lowell Lectures, 1926*. Cambridge: Cambridge University Press, 1927.

———. *Science and the Modern World*. Cambridge University Press, 1925.

———. *The Function of Reason*. Princeton, NJ: Princeton University Press, 1929.

Williams, Daniel Day. "The Prophetic Dimension." In *The Uniqueness of Man*, edited by John D. Roslansky, 139–63. Amsterdam: North-Holland, 1969.

Williams, Patricia A. *Doing Without Adam and Eve: Sociobiology and Original Sin*. Theology and the Sciences. Minneapolis, MN: Fortress Press, 2001.

———. "How Evil Entered the World: An Exploration Through Deep Time." In *The Evolution of Evil*, edited by Gaymon Bennett, Martinez J. Hewlett, Ted Peters, and Robert John Rossell, 204–17. Religion, Theologie und Naturwissenschaft 8. Göttingen: Vandenhoeck and Ruprecht, 2008.

———. "Sociobiology and Original Sin." *Zygon* 35, no. 4 (2000): 783–812.

Williams, Raymond. "Social Darwinism." In *Herbert Spencer: Critical Assessments*, edited by John Offer, 2:186–97. New York: Routledge, 2000.

Wilson, Edward O. *Sociobiology: The New Synthesis*. Cambridge, MA: Harvard University Press, 2000.

Winchell, Alexander. *Preadamites: Or a Demonstration of the Existence of Men before Adam, Together with a Study of Their Condition, Antiquity, Racial Affinities, and Progressive Dispersion over the Earth*. Chicago: S.C. Griggs, 1880.

Wolpoff, Milford H., and Rachel Caspari. *Race and Human Evolution*. New York: Simon and Schuster, 1997.

WORKS CITED

Woods, Henry. *Augustine and Evolution: A Study in the Saint's* De Genesi ad Litteram *and* De Trinitate. New York: The Universal Knowledge Foundation, 1924.

Wright, Chauncey. *Darwinism: Being an Examination of Mr. St. George Mivart's "Genesis of Species."* London: John Murray, 1871.

Yang, Chin Jian, Luis Fernando Samayoa, Peter J. Bradbury, Bode A. Olukolu, Wei Xue, Alessandra M. York, Michael R. Tuholski, et al. "The Genetic Architecture of Teosinte Catalyzed and Constrained Maize Domestication." *Proceedings of the National Academy of Sciences* 116, no. 12 (March 19, 2019): 5643–5652.

Žižek, Slavoj. *The Indivisible Remainder: An Essay on Schelling and Related Matters*. London: Verso, 1996.

Index

abstraction, 59, 87, 100–101, 105, 127, 209, 215
actual entity, 101–2, 104–6
actual occasion, 101–6, 113–14, 202
actual sin, 151–54, 156, 159, 180
Adam and Eve, 64, 121–22, 125, 133–34, 144, 149. *See also* fall of humanity; pre-Adamitism
adaptation, xiv, 4, 19, 20n33, 22–25, 31–32, 46–48, 50, 53, 165, 168, 177, 194, 213
agonism, 24, 61, 63, 83, 85, 94
Alexander, Samuel, 80, 95
altruism, 32, 166, 170, 175, 190–91
Ambrose of Milan, 149
Anaximander, 4n4
and superject, 103
Anselm of Canterbury, 151
anthropocentrism, xviii, 11, 28–29, 35–37, 39–40, 44–45, 89, 201–4
anthropology, 133, 137–39
apocalyptic, 115, 206
Aquinas, Thomas, 11, 57, 62, 151–52, 168n47, 190. *See also* Thomism
on original sin, 62, 151–52
Arian controversy, 149
Aristotle, 5–8, 12–13, 57, 60n41, 82, 93, 136
categories of, 93, 132
god of, 112, 114, 209

on form and matter, 100, 126–28, 130–33
on substance versus accidents, 5, 100–101, 109
Articulo, Archimedes C., 194
atheism, xiv, 21, 41–42, 50, 164, 166
Aufhebung. *See* sublation
Augustine, 8–11, 57, 107, 110, 140, 170, 176, 190, 209, 212
on original sin, 62, 68, 148–51, 159
autonomy, xiv, 20–21, 61, 83, 123, 182
Ayala, Francisco, 28, 144–45

Bacon, Francis, 99–100
Baltazar, Eulalio, 108
Balthasar, Hans Urs von, 186
Barbour, Ian, 63, 99
beauty, ix, xv, 2, 5–7, 14, 43–46, 55–56, 58, 90, 97, 111, 203, 205, 208
Benedict XVI, 36, 124
Berdyaev, Nikolai, 39
Bergson, Henri, 50, 52–53, 56, 63, 91–92, 98, 131
élan vital, 53, 92
Bible, xii, xvi, 12, 26, 72, 112n88, 121, 153, 190, 199, 208–9
1 Corinthians, 45, 87
1 John, 110
Colossians, 72–73
Exodus, 110

INDEX

Genesis, 9, 29, 52, 88, 117–18, 133–34, 137, 140, 144, 148, 163, 167, 199, 202
John, 72, 88–89
on God, 110, 163, 208–9
Paul, 45, 72, 136, 148
Philippians, 72
Revelation, 71
Romans, 171
Wisdom, 148
Big Bang, 96
Birch, Charles, 201
Blumenbach, Friedrich, 137–38
body, 8, 33, 45–46, 59–60, 89, 94, 104, 121, 123–31, 140, 149. *See also* organicism
as corpse, 171–72
as expression of soul, 90, 129
as machine, 48–49
resurrection of, 128, 193
Bonaparte, Napoleon, 74
Bonaventure, 9
Bonifazi, Conrad, 181–82, 200–201
Bosanquet, Bernard, 95
Bostrom, Nick, 194
Boveri, Heinrich, 47
Bowler, Peter J., 5n4, 17n27, 20, 49
Brachtendorf, Johannes, 159n16
Bradley, F. H., 80, 95, 98
brain, 29, 45, 58, 60, 129–30, 177
Brandi, Salvatore, 120
Bresch, Carsten, 177
British idealism, 62, 80, 95, 98
Bruteau, Beatrice, 108
Bryan, William Jennings, 27
Bulgakov, Sergei, 69–70

Caesar, Julius, 186
Calvin, Jean, 154, 161n22
Cann, Rebecca L., 143

capitalism, 33, 173, 196, 200
Chambers, Robert, 21
chaos, 2, 4n4, 22, 24–25, 29, 31, 94, 110, 157, 185–86, 202
charity, 33, 174, 191
Chicago School of, 108
Christ, xv, 41, 45, 55, 66–67, 69, 71–74, 115, 140–42, 148, 178, 182, 209, 215
crucifixion of, 67, 87, 161–62, 170
God-forsakenness of, 161
imitation of, 170, 174, 191
incarnation of, xv, 9, 55, 71–73, 89, 107, 115, 162, 182, 209, 215–16
resurrection of, 148
Christocentrism, 41, 55, 71–72, 75, 115, 182, 215
Church, xvii, 45, 61, 70, 113
as body of Christ, 45
Catholic, xii, xvii, xix, 40–41, 56–57, 60–64, 67–72, 74, 77–78, 108, 115, 123–25, 134, 137, 140–43, 149–52, 154, 157–58, 162, 175
in relation to the state, 61, 83
Magisterium of, 117–23, 145–46. *See also Humani generis*
Orthodox, 70, 108, 150n3
civilization, 31–32, 113
climate change, 197, 199
Cobb, John Jr., 108, 201, 203
Cole-Turner, Ronald, 196–97
colonialism, 134–36, 138, 187–88, 199–200
Comenius, John Amos, 13n18
competition, 22, 24, 27–28, 31–32, 35, 155, 161–62, 165–67, 169, 177, 187–88, 203, 205
concrescence, 93, 102–5, 114

INDEX

concupiscence, 122, 149–50, 152, 154, 188, 191, 196
contextually conscious forms of theology, 211
conversion, 74, 153–54, 180, 197, 213
Coon, Charleton S., 138–39
Cope, Edward Drinker, 47
Copernicus, Nicolaus, 28
cosmology, 14, 93, 112, 204. *See also* world
Council of Cologne, 140
Cousins, Ewert, 78n1, 108
Cowdin, Daniel, 200n45
creation of woman, 29
creation out of nothing, 52
creativity, 49–54, 58, 63, 77, 131, 182, 184, 186, 194, 205
   in Whitehead, 91–95, 97–99, 102–7, 109–10, 114, 182, 208
   of God, 43, 52, 68, 110, 112, 116, 132–33, 161, 168, 172, 190
Crusafont, Miquel, 118
Crysdale, Cynthia, 188–89, 215
culture, x, xiii, xix, 2–3, 5, 12, 16, 22–23, 26, 29, 32, 34–35, 42, 59, 67, 78, 85, 134–36, 142, 164, 166–67, 169, 174–75, 177–78, 180, 183–84, 187–89, 191–93, 202, 205
Cuvier, Georges, 13

Daly, Gabriel, 191–92
Dana, James Dwight, 26n53
Darwin, Charles, 3–4, 11, 13–36, 39, 41–43, 46–50, 54, 56, 74, 79, 109, 119, 138–40, 155, 203
   on theodicy, x–xi, 21–22, 147
Darwinism, 5, 26, 28, 47, 51–52, 97, 119, 138
   social, 30–35, 42
Dawkins, Richard, xiv, 165–67, 187

de Lubac, Henri, 65n57, 67–68, 73
Deane-Drummond, Celia, 177–78, 183, 186–87, 193
death, 20, 22, 25, 27, 31, 65–67, 87, 110, 128, 148, 155, 160–61, 164, 168–69, 171–73, 193, 202–3
   and decomposition, 171–72
deforestation, 197
Deism, 43, 114
Delio, Ilia, 68, 70–71, 108, 194–95
Dennett, Daniel, xiv
Depoortere, Frederiek, 161n22
Descartes, René, 61, 98, 100, 104
determinism, 52, 92, 112–13, 115, 182, 184
   biological, 34, 166, 187
dialectic, xviii, 24, 62, 77, 79, 81, 83–95, 97, 99, 101, 103, 105–7, 109, 111, 113, 115, 155, 164, 184–87, 189, 208
Diderot, Denis, 5n4
Dobzhansky, Theodosius, 139
dogmatism, xii, xviii, 12, 41, 68, 124
Domning, Daryl, 166
Dorrien, Gary J., 79
drama, xv, 28, 73, 148, 181–83, 186
dualism, 60–61, 123, 126, 195
dualities, 61, 63, 83, 85, 94, 104, 126–28
dynamism, 23, 43, 54, 62, 91

Eckhart, Meister, 57, 62
ecology, xv, xix, 37, 101, 183, 197–205
ecotheology, 198–204
Edwards, Denis, 191, 201
Einstein, Albert, 74, 126, 211
electricity, 53, 126, 129
Elijah, 68
Ellis, Robert, 80
Empedocles, 4–5, 24n49
energy, 66, 72, 98
   in physics, x, 103, 126

239

INDEX

Enlightenment, 3, 5, 61, 154
Entscheidung, 157
eschatology, 9, 115, 182, 193–94, 197, 216
eternal life, 193
ethics, 23, 34, 188–89, 196, 200, 202–5
eugenics, 34
Eve hypothesis, 143–44
evil. *See* theodicy
evolutionism, xviii, 2–4, 16, 26, 35–36, 56, 74, 79, 117, 158, 181, 198, 210
experience, xiii, 2, 18, 36, 56, 91, 97–99, 114, 126–27, 129, 164, 167, 174, 176, 178, 190–91, 194, 197, 211–12
exploitation, 34, 179, 199–200
Exsultet, 162
extinction, 5, 19, 24, 31, 39, 197, 204
extraterrestrials, 205

faith, xiii–xv, xvii, 41, 61, 63, 66, 71, 83, 85, 109–10, 112, 123, 125, 141, 145, 150, 152–53, 212, 214, 216
fall of humanity, 149, 151–52, 154, 158–60, 162, 165, 176. *See also* Adam and Eve
    in Hegel, 163–64
    inverted, 160, 162–64, 167, 169, 176
fate, 74, 93, 112
Father, 86–88, 161, 209, 214, 216
feminist theology, xvi–xvii, 211
feudalism, 135, 199
Feuerbach, Ludwig, 90
Fichte, Johann Gottlieb, 79, 83–84
Filaret of Moscow, 150n3
Fiske, John, 52, 163–64
Fleming, Walther, 47n14
Foley, Leo, xvi
forgiveness, 129, 190

Foure, Elie, 192
free-will, 62, 149–50, 152, 156–57, 167, 176
    as consequential, 172
freedom. *See* openness
Freud, Sigmund, xiii
fundamentalism, xii

Galilei, Galileo, 11–12, 74
Galton, Francis, 34
gene editing, 192, 195
generation, 131–33
genetics, 122, 129, 139–40, 143–45, 155, 158, 165–67, 177, 212. *See also* gene editing
genius, 74, 97, 148, 186
genocide, 33–34, 67
geology, 18, 54, 71, 132
giddiness, 84
Gilbert, W. S., 51
Gilson, Étienne, 7n8, 112
Girard, René, 179–80
globalization, 192, 200
Gnosticism, 89, 158, 160, 162–63
God, ix–xi, xiii–xv, xvii, xix, 6–14, 20–22, 24, 29–30, 36–37, 42–44, 49–50, 52, 54–55, 61–62, 66–73, 75, 77–79, 81–83, 86–95, 97–98, 101, 104–7, 109–15, 117–18, 121, 123, 125–26, 128, 131–33, 146–49, 151–55, 157, 159–64, 167–72, 174–76, 178–80, 182, 189–95, 197, 199, 202, 204, 206–9, 212–16. *See also* Trinity; Father; Christ; Holy Spirit
    as Creator, ix, xi, 9, 11, 13, 15, 42–43, 52, 68, 70–72, 110, 112, 116, 125, 132–33, 161, 163, 168, 172, 179, 190–91
    as Prime Mover, 112, 114, 191

as spiritual, 88–89, 212. *See also*
    Hegel, Georg Wilhelm
    Friedrich
death or suffering of, 87, 94, 161n22,
    174, 208, 214–15
eternity of, 82, 89, 112, 114, 132
immanence of, 42–43, 70, 75, 87, 94,
    107, 112, 208–9, 216
in Aristotle, 112, 114, 209
in classical theism, 209. *See also*
    Aristotle
in the Bible, 110, 163, 208–9
intervention of, xiv, 12, 14, 20–21,
    43–44, 92, 111–14, 132–33,
    174, 190–91, 205, 208–9
omniscience of, 43
power of, x, 21, 70, 133, 155, 160,
    189–90, 208
self-revelation of, 42, 72–73, 77, 82,
    86, 89–90, 117, 160, 190, 204
transcendence of, x, 42, 70, 72, 86,
    89, 107, 128, 131–32, 190–91,
    208–9
Gospel, xv, xvii, 214–15
Gould, Stephen Jay, 5, 25
grace, 9, 133, 148, 151–53, 161, 170,
    174–75, 178–80, 189–92, 194,
    205–6, 214–16
    and free-will, 62, 149, 152, 189
    gratuity of, 69, 89, 107, 148, 153, 191
Gray, Asa, 26n53, 43
Great Chain of Being. *See*
    scala naturae
Gregersen, Neils Henrik, 161–62
Gregory of Nazianzus, 150n3
Grumett, David, 70
guilt, 148, 150n3, 151–54, 169, 179–80,
    190
Gustafson, James, 204

Haeckel, Ernst, 23
Haight, Roger, 190
harmony, xiii, 2, 11, 46, 56, 97, 117, 202,
    205
Harnack, Adolf von, 154
Hartshorne, Charles, 108
Haught, John, xiv, 21, 168, 205
Hawks, John, 143n43
Hefner, Philip, 114, 166–67, 183, 194
Hegel, Georg Wilhelm Friedrich, xviii,
    37, 40, 78–81, 95, 97–98, 104,
    106–8, 162
    on dialectic, 62, 81–88, 92–95, 103,
        164, 208
    on dialectical history, xviii, 81,
        184–86
    on God–world relationship, 81–82,
        86–95, 107, 164, 208
    on providence, 111–13, 185
    on Spirit, 87–90
    on the fall, 163–64
    on the state, 185–86
    on theodicy, 107, 110, 155, 164, 208
hell, 151, 153, 159–60, 172
Hick, John, 159–60, 162, 164, 174, 189
Hinduism, 91
historicity, 134, 142, 148, 152, 157, 172,
    175–79, 215–16
history, xii–xiii, xvi–xvii, 1–5, 9,
    11–12, 15, 18, 26–27, 49, 53–56,
    58, 60, 62, 65–66, 72, 74, 79,
    82, 84, 87–88, 91–93, 98, 106,
    110–11, 114–15, 119, 123–25,
    128, 134–35, 140, 142–43, 148,
    151–52, 158–59, 161, 163, 172,
    176, 178–87, 195, 199, 215–16
    dialectical, xviii, 81, 184–86
    of salvation, 113, 180, 192, 205
Hitler, Adolf, 33

INDEX

Holy Spirit, xix, 86–87, 209, 214, 216
Homo erectus, 122, 133, 138–39, 143–44
hope, 59, 64, 67, 148, 159, 175, 181, 186, 196–97, 216
horse, evolution of, 17, 50, 131–32
humanity, xii, xviii–xix, 8, 11, 18, 22, 25–26, 28–31, 35, 39–41, 44–45, 52, 58–60, 62, 65–67, 72, 74, 82, 90, 118, 124, 128, 133, 136–41, 144, 146–48, 151–52, 154, 160–61, 163–65, 169, 171, 176, 178, 183, 186, 192, 194–99, 201, 203, 206, 211. *See also* anthropocentrism
    as created co-creators, 79, 114, 169, 194. *See also* Hefner, Philip
    dignity of, xii, xviii, 27, 58, 73, 118, 137, 195
    emotions of, 129–30
    evolution of, xiv, 25, 27–29, 60, 64, 74, 117–18, 121–24, 129–31, 133–47, 150, 163, 167, 176–78, 188, 196
    teleology of, 29, 142, 178
    unity of, 58–59, 73, 122–23, 138, 140–44, 165, 178–79, 215–16
Hume, David, 95–96
Hus, Jan, 150
Huxley, Julian, 47, 181, 194
Huxley, T.H., 50–51, 53
Hyatt, Alphaeus, 47
hypostatic union, 72

idealism, 95, 97, 106
Ideas, 5, 13, 15, 39, 86, 104, 111–14, 213
    as manifested in the world, 5–7, 45, 58, 90, 97, 99, 106, 111, 127
identity, 85, 91, 102, 136, 146, 197

ideology, xiv, xviii, 2, 16, 31–33, 56, 134–35, 194
Illingworth, J. R., xi, 1, 44–46
Index of Prohibited Books, 120
individualism, 34, 53, 74
individuality, 45, 49, 58–59, 61–62, 73–74, 83, 85, 87, 99, 113–14, 125, 128, 180, 184–86, 188, 204
industrialization, 196–97, 199–200
inequality, 11, 33, 65, 137–38, 173, 196
instinct, 155, 191
intellect, 127
involution, 44
Irenaeus of Lyon, 159–60, 162n29
irony, 6, 83, 85–86, 88, 110, 162–63

James, William, 98
Jesus. *See* Christ
Jewish tradition, 9, 89, 148
John of Damascus, 61
Jonas, Hans, 163n29
justification, 110

Kant, Immanuel, 79, 93
    on noumena and phenomena, 61, 96–97
    on radical evil, 156–57
    on reason and faith, 61, 83, 97, 109
    on theodicy, 109
    on transcendental dialectic, 84
Kierkegaard, Søren, 130
King, Ursula, 68
Kingsley, Charles, 43
Korsmeyer, Jerry, xix, 166, 205
Kurzweil, Ray, 194

La Peyrère, Isaac, 136–37, 140
LaCugna, Catherine Mowry, 69n70
Lamarck Jean-Baptiste, 21–22, 46–50, 53

law, xi, 10, 12–13, 19, 30, 33, 43, 47, 70, 112, 186, 200n44, 210
Leibniz, Gottfried Wilhelm, 22, 46, 79
  on theodicy, 13–15, 109–10, 155–56, 159
Leonardi, Piero, 118
Leopold, Aldo, 204
Leroy, Dalmace, 120
leveling of differences, 59
Libertarianism, 173
Linnaeus, Carl, 13
Locke, John, 61, 95
logocentricism, 98
logos, 58, 98
love, ix, xix, 5, 30, 51, 59, 66, 69, 74, 77, 86, 89, 107, 129, 154, 170, 179, 202, 204–5, 209, 215
Lovejoy, Alfred O., 8
Lucas, George R. Jr., 79n4, 80, 92–93
Lukas, Mary and Ellen, 60n41
Luther, Martin, 61, 110n82, 161
  on original sin, 150–54

magnetism, 99–100, 126
Mahoney, Jack, 166, 175, 190–91
Malthus, Thomas, 17, 27–28, 33–34, 172–73
Manichees, 89, 110, 148
Marcion, 135
Maritain, Jacques, 208
materialism, x, xii, xiv, 32, 41–43, 48–51, 60–61, 95–96, 106, 145
mathematics, 12, 23, 52, 56, 96–97, 145, 173
matter, 12, 24, 49, 94, 104, 121, 124–25, 128, 132–33, 172
  as pre-history of spirit, 62, 124, 129–31
  in Aristotle, 100, 126–28, 130–33
  in Plato, 125–28
  in science, x, 126, 210, 212
  in Teilhard, 55, 60–63, 71
Maupertuis, Pierre Louis Moreau de, 49
McFague, Sallie, 202–4
McGrath, Alister, 20n33
McNally, David, 173
meaning, ix–x, xii, xiv–xv, 2–3, 9, 11, 20, 24–27, 29–31, 35–37, 41–46, 48–49, 53–54, 56–58, 60–64, 66, 72–73, 77, 81–82, 88, 91–92, 94–99, 105–6, 113–15, 124, 126–29, 131–32, 147, 172, 174, 181–82, 184–85, 194, 200–202, 212–13, 215
mechanistic thinking, 6, 11–12, 14, 19, 44–45, 49–54, 57, 67, 70, 92, 98, 131, 186
mechanization of production, 192, 196, 200
medieval theology, 3, 8–12, 101, 110, 112, 150, 199–200, 209. *See also* scholasticism
Meiners, Christoph, 137–38
Meléndez, Bermudo, 118
Mellert, Robert B., 108
Mendel, Gregor, 47, 212
mercy, 107, 148, 153–54
merit, 151–52
Mesle, C. Robert, 93
metanarrative, 2–4, 9, 11, 20, 22, 25–26, 35–36, 56, 78, 210
metaphor, 156, 166, 210, 213
metaphysics, xvii, 14–15, 21, 46, 48–50, 55, 60, 63–64, 68–69, 71, 79, 81, 84, 91, 93, 95–99, 101, 103–4, 110, 131, 133, 141, 207–8, 210, 213
Metz, Johann Baptist, 214–15
Midgley, Mary, xiii, 204
Milton, John, xi, 109–11, 159, 213–14

INDEX

Mivart, St. George Jackson, 25n52, 119–20, 124
modern synthesis, 47, 50, 139
modernist crisis, 64, 118
modernity, 3, 61, 97, 109, 135, 154, 158, 168
Moltmann, Jürgen, 161, 202, 214
monism, 52. *See also* Leibniz, Gottfried Wilhelm
monogenism, 133–35, 137–45
monophyletism, 140, 142–44
Moore, Aubrey, 26n53
morality, x, xvii, xix, 14, 31, 33, 59, 64, 66, 81, 135, 155–56, 159, 163–66, 168–69, 171, 173, 176, 179, 183–84, 187–89, 191, 193, 195–97, 200, 203–5, 213–14, 216
multiregional theory, 143–44
myth, 72, 105, 111, 214

National Socialism, 65
natural selection, x, 3–5, 10–11, 13–14, 16–29, 31–32, 36–37, 39, 42–43, 46–55, 73, 78–79, 117, 119, 121, 131, 134–35, 138–39, 155, 161, 164–65, 168–69, 172, 183, 188, 192, 201, 213. *See also* Darwin, Charles
  and chance or randomness, 17, 20, 51, 194, 212
  and niche construction, 177
  and time, 3, 17–19, 60, 177
  as dysteleological, 21–22, 36, 41–42, 91–92, 97, 133
  as struggle for existence, 19, 28, 35, 46, 55, 63, 165
natural theology, xviii, 42–44, 54–55
nature does not make a leap, 7, 12–13
Nemesius of Emesa, 8
neo-Darwinism. *See* modern synthesis

neo-Lamarckism, 47–48
neo-scholasticism, 57
Neoplatonism, 5, 8, 14, 45, 57, 90, 97, 112
newness, 52–53, 94, 131–32
Newton, Isaac, 12–13, 49, 74, 96
Nietzsche, Friedrich, 35, 79, 208
Nisbet, Ebenezer, 51
Nogar, Raymond J., 2n3
nonduality, 60, 62
noosphere, 54, 59–60, 63, 65, 71, 73–74, 193
Northcott, Michael, 200n45

O'Regan, Cyril, 86, 163
Ogden, Schubert M., 108
Olkovich, Nicholas, 166
ontology, 81, 91, 100–102, 104, 107, 109, 209n3
ontotheology, 208
openness, 43, 48, 52–53, 92, 184–89. *See also* creativity
optimism, 30–31, 43, 55, 64–66, 161, 171, 175, 182–83, 192–94, 201. *See also* hope
organicism, 15, 41–42, 44–46, 54–59, 73, 75, 201, 204
Origen, 89
original sin, xix, 111, 116, 122–23, 154–59, 162, 164, 168, 170, 172, 174–78, 196, 216. *See also* actual sin; fall of humanity; guilt; historicity
  and polygenism, 122–23, 133, 140–42, 178
  and theodicy, 159, 175–76, 179
  Aquinas on, 62, 151–52
  as a doctrine of grace, 148, 174, 191, 214
  as codetermination of freedom, 178–80

244

as inherited guilt, 150n3, 151–52
as loss of grace, 62, 151–52, 178–79
Augustine on, 62, 68, 148–51, 159
in Calvin, 154
in Luther, 150–54
in selfish gene theology, 155, 167
propagation of, 142, 170, 174–75, 178. *See also* historicity
Teilhard on, 64, 118
Ormerod, Neil, 188–89, 215
Orr, James, xvi, 44
Out of Africa hypothesis, 140, 143–44
overpopulation, 173

paganism, 5, 7n8, 9, 72, 111–13, 186, 212
paleontology, 54, 56, 118, 139
Paley, William, 42–43
pantheism, 91, 112
paradox, 70, 85, 107, 209
parasites, xi, 21–22
pathos, 98
patriarchy, xvii
patristic theology, 10, 52, 89, 110, 150n3, 209
peace, 24–25, 65, 179, 183
Peacocke, Arthur, 63
Pelagianism, 149, 151, 170–71, 175, 205
Pernoud, Régine, 200n44
persecution, 67
pesticides, 189
Peterson, Gregory, 164, 166–67
Philip the Handsome, 200
physics, 2, 12, 23, 49, 56, 75, 82, 96, 103, 177, 213. *See also* energy
Pius XII, 117
Humani generis, xix, 119–24, 129, 133–34, 139–42, 145
Plato, 4–8, 10–13, 15, 22, 45–46, 49, 57–58, 84, 89–90, 97, 106n68. *See also* Ideas

on matter and spirit, 125–28
on providence, 111–13
Plotinus, 5n5, 45, 57–58, 90
politics, 2, 12, 16, 25, 31, 33, 61, 80, 118–19, 134–36, 138, 172–73, 180, 188, 193, 200, 216
Polkinghorne, John, 63, 75
polygenism, 121–23, 133–38, 140–42
polyphyletism, 122, 133–34, 138–44
positivism, 145
postmodernity, 98, 164, 193, 197
poverty, 32, 173, 200
praxis, 182, 197–98
pre-Adamitism, 136–37, 140–42
prehension, 99–106
prehuman hominins, 56, 122, 139, 143, 167
pride, 152, 173, 202
process theology, 36–37, 55, 70–71, 78n, 95, 109, 157, 160, 164, 168–69, 174, 194, 201, 207–9, 213. *See also* Birch, Charles; Cobb, John Jr.; Ogden, Schubert M.; Whitehead, Alfred North; Williams, Daniel Day
Proclus, 90
Protestantism, xii, 12, 42, 119, 147, 150, 152–54, 157–58, 163, 166, 168, 200
providence, x–xi, xvii, 9–15, 42–44, 73, 109, 189
in Hegel, 111–13, 185
in Plato, 111–13
in Whitehead, 111, 113–15, 174, 185
Pseudo-Dionysius, 57
psychology, xv, 23, 37, 98, 164
Pythagoras, 97

qualitative leap, 130, 132
quantitative reckoning, 67, 121, 130, 172–73
quantum entanglement, 210, 213

racism, 22, 31, 33–34, 134–41, 144
Rahner, Karl, xix, 63, 117, 119, 121–25, 128–33, 141–42, 145, 148, 176, 190–91, 195
   on the codetermination of freedom, 178–80
Ratzinger, Joseph. *See* Benedict XVI
realism, 106
reason, xv, 14, 58, 61, 83–85, 96–98, 109, 113, 152, 185
reconciliation, dialectical, 83–85, 87–88, 93–94, 185
Reformation, Protestant, 3, 12, 61, 147, 150
relationality, 100–102, 178, 202–4, 210
religion, 9, 26, 32, 71–72, 78, 111, 184, 186, 194, 212. *See also* science
Renaissance, 200
right to use nature, 199, 200n44
Romanticism, 98, 203
Royer, Clémence, 32–34
Ruether, Rosemary Radford, 202–3

salvation, 61, 79, 140–42, 153, 161, 167, 169–71, 174, 181, 214
   as gift, 148–49, 151, 171, 175
   history of, 113, 180, 192, 205
Santmire, H. Paul, 200–1
Satan, 110
Savage, Minot J., 159
scala naturae, 7–11, 22–23, 25, 28
scapegoat, 179
Schelling, Friedrich Wilhelm Joseph von, 5n4, 61, 79, 106, 157
Schleiermacher, Friedrich, 160, 190
Schmitz-Moormann, Karl, 40n4, 68
scholasticism, 93, 101, 112n88
Schoonenberg, Piet, 69n70
Schwager, Raymund, 147–48, 176–80

science, 5–6, 12, 16, 32, 34, 39, 56, 96–97, 133–35, 144, 197. *See also* genetics; natural selection; sociobiology; technology
   and religion/theology, ix–xix, 11–12, 20–23, 27, 29, 36, 40–41, 43, 54, 64, 72, 75, 77, 83, 97, 117–24, 145–46, 148, 158, 176–77, 207, 210–13
   as authority, xiv, 2–3, 36, 121, 145–46, 202–3, 210–11, 213
   as hermeneutical tool, 211
   empirical, x, xiv, 2–4, 15–16, 18, 21–24, 26, 36, 42–43, 48, 56, 64, 133–35, 139, 142, 166, 213
   on matter, x, 126, 210, 212
   opportunistic appeals to, xiv, 36, 202–3, 210–11, 213
segregation, 180
selfish gene theology, 165–67, 169
sense-perception, 7, 99, 126–27
Seuss, Eduard, 59n38
sexual selection, 25, 48
Sheen, Fulton, 210
Sideris, Lisa H., 202–5
simplicity, 14
Simpson, Richard, 21, 168n47
slavery, 62, 136, 138, 149, 196, 208
social sin, 179–80, 188–89
socialism, 31
sociobiology, 37, 165–66, 187
sociology, 23, 167, 211, 213
Socrates, 84
Solovyov, Vladimir, 28
soul, 59–60, 62, 126–30
   and soul-making, 159–60
   as subsistent form, 128
   body as expression of, 90, 129
   immediate creation of, 27, 121–25, 190–91

immortality of, 27, 128
    in relation to God, 89, 212
Spencer, Herbert, 22–24, 31–33, 46, 52
Spinoza, Baruch, 91
Stoicism, 8, 89, 112
Stoneking, Mark, 143
subjectivity, 59, 83–84, 112, 154, 185, 202, 204
sublation, 87–88, 92–93, 106, 113. *See also* reconciliation, dialectical
substance, 78, 100–1, 104, 126
    versus accidents, 5, 28, 50, 53, 100–1
Sullivan, Arthur, 51
Sutton, Walther, 47n14

technology, xii, xix, 1–2, 132, 183, 192–97, 199, 205, 216
Teilhard de Chardin, Pierre, xii, xv, xvii–xviii, 36, 39–41, 54–61, 64–75, 77–78, 98, 108–9, 114–15, 117, 122, 124, 131, 138–40, 171, 182–83, 192–94, 201–2, 205, 215
    censorship of, 64, 118n2
    on matter, 55, 60–63, 71
    on original sin, 64, 118
    on the atomic bomb, 65–66, 192–93
    on theodicy, 66–68, 70–71, 108–9
    personalism of, 58–59, 73–75
teleology, 10, 15, 20, 22–23, 25, 42, 44, 53–54, 56, 97, 102, 104, 113, 115, 118, 131, 133, 202, 210
    of humanity, 29, 142, 178
    of the world as a whole, 58, 91
temporality, 1–2, 9–10, 29–30, 58, 68, 73, 89, 91, 104, 106–7, 132, 181–82, 216
    in natural selection, 3, 17–19, 60, 177
Tennyson, Alfred, 30

Tertullian, 135
theodicy, 37, 157, 160–64, 166, 168–69, 171–74, 207, 213–15
    and convertibility of good and evil, 67, 171–72
    and cosmic immaturity or inertia, 64–65, 158–60, 169, 174, 176
    and economic logic, 33, 35, 67, 69, 172–73
    and natural evil, 20, 66, 155, 161, 168–69, 171, 176, 179, 214
    and original sin, 159, 175–76, 179
    and the inevitability of evil, 65–66, 98, 155, 159, 169–70, 174, 214
    in Bible and tradition, 67, 110, 159, 168, 176, 179, 209
    in Darwin, x–xi, 21–22, 147
    in Hegel, 107, 110, 155, 164, 208
    in Kant, 109
    in Leibniz, 13–15, 109–10, 155–56, 159
    in Metz, 214–15
    in Teilhard, 66–68, 70–71, 108–9
    in Whitehead, 70–71, 107, 109–11, 147, 160, 208, 214
Thomism, 57, 60, 63, 124–28
Tissot, Jacques, 12n18
Toolan, David, 201
tradition, xv–xvii, 8, 12, 29, 36, 41, 55, 57, 60–63, 67, 69–70, 72, 75, 86, 89, 95, 107, 112n88, 117, 122–23, 125–26, 133, 139–42, 146, 148–49, 152, 154, 156, 158–59, 162–63, 168, 170, 172, 174–76, 199, 208–9, 212–13, 215
    development of, xviii, 9
transcendental decision, 156–57
transcendentals, 93
transhumanism, 194, 196–97

Trinity, 69, 86–88, 178, 191, 209, 211, 214, 216
typological thinking, 5–6, 15

ultrahumanism, 194–95
undifferentiation. *See* leveling of differences
Ussher, James, 18
utilitarianism, 44–45, 200

Vatican Council II, xvii, 119
Vernadsky, Vladimir, 59n38
violence, 14, 24–25, 31, 63–64, 66–67, 110, 134, 155, 165, 167–68, 177, 179–80, 187, 199–200, 203, 205
vitalism, 41–42, 46, 49–50, 52–54
Vogt, Karl, 138
vulnerability, 167

Wallace, Alfred Russel, 3–4, 11, 15–16, 24, 29, 74, 119, 155
Weinandy, Thomas G., 69n70, 209n3
Welch, Sharon, 188
White, Lynn Jr., 199–200, 204
Whitehead, Alfred North, xviii, 79–80, 103–4, 215. *See also* actual entity; actual occasion; concrescence; prehension
  atomism of, 103
  legacy of, xii, 75, 78, 81, 108–9, 114–15, 201
  on creativity, 91–95, 97–99, 102–7, 109–10, 114, 182, 208
  on feeling, 98–99, 102, 114, 174, 202–3
  on God–world relationship, xvii, 70–71, 81, 92–95, 98, 104–15, 174, 202, 204, 208
  on history, xviii, 81, 98, 184–86
  on meaning, 95–98
  on providence, 111, 113–15, 174, 185
  on theodicy, 70–71, 107, 109–11, 147, 160, 208, 214
Williams, Daniel Day, 39–40, 108
Williams, Patricia A., 166, 170–72, 190
Wilson, Allan C., 143
Wilson, Edward O., 165–66
Winchell, Alexander, 137
Wolpoff, Milford. *See* multiregional theory
wonder, ix, xix, 46
world, xi–xv, xvii, 1–15, 18–19, 21–23, 26, 30, 35–36, 39–46, 48–50, 52–54, 56–62, 64–65, 83–84, 96–97, 99–101, 103–4, 127, 142, 147, 152, 155–57, 169, 171–74, 179–83, 185, 187–88, 195, 197–205, 211, 214–16
  and diversity, 6, 11, 14, 17, 19, 90–91, 94, 197
  in relation to God, ix–x, xvii, 54, 66–75, 77–79, 81–82, 86–95, 98, 104–7, 117–18, 128, 131–33, 146, 155, 159–64, 168, 174, 176, 189–91, 197, 199, 202, 204, 208–9
  teleology of, 58, 91
  unity of, 6–7, 15, 45–46, 55–59, 63, 73, 201–2
Wright, George Frederick, 26n53
Wycliffe, John, 150

Zahm, John, 120